ST DA - X Y -

AN INTRODUCTION TO THE ANALYSIS AND PROCESSING OF SIGNALS

AN INTRODUCTION TO
THE ANALYSIS AND PROCESSING OF SIGNALS

PAUL A. LYNN

Reader in Electronic Engineering
University of Bristol

Second Edition

M

First edition 1973
Reprinted 1977, 1979, 1980
Second edition 1982

Published by
THE MACMILLAN PRESS LTD
London and Basingstoke
Companies and representatives
throughout the world

Printed in Hong Kong

ISBN 0 333 32994 5 (hard cover)
ISBN 0 333 34030 2 (paper cover)

Contents

PART II SIGNAL PROCESSING

Preface

The aim of this book is to provide a concise and readable introduction to the theory of signal analysis and linear signal processing. Its level should make it suitable for the second and final years of degree courses in electrical and electronic engineering, and for other courses which deal with the analysis and processing of signals, records, and data of all types. A small proportion of the material covered would normally be reserved for courses of postgraduate standard.

The author of a book on signal theory—a subject already covered by a number of recent texts—must presumably offer some excuse for writing it. In my case, I have for some time felt the need for a book in this general area which neither assumes, nor develops, any detailed background in electrical circuits or techniques. The approach I have adopted is to discuss signals and signal processing from what is generally called a 'systems' viewpoint: in other words, the text does not involve itself with the details of electrical or electronic signal processors, nor does it concentrate on those types of signal which are largely or exclusively the concern of electrical and electronic engineering. I hope and believe, however, that students of electrical engineering will find the text relevant to courses on communications and linear systems as well as on signal theory, and that it will help them to fit much of their work on electric circuits and systems into a more general framework. I also trust that the approach adopted will make the text useful to workers in other fields where signals and data are important, such as other branches of engineering, physics, medicine and certain areas of social science and economics. With the latter in mind, I have tried to keep the mathematics as straightforward as possible, and to assume knowledge of nothing more advanced than basic integral calculus and the elementary manipulation of complex numbers. A further advantage of omitting detailed work on electrical circuits and systems is that it makes the task of describing basic aspects of signal theory in a short book that much more manageable.

I hope it may not seem unreasonable if I claim some novelty in content as well as in general approach. I have paid rather more attention to sampled-data signals than is usual in a book of this length, for several reasons. Electronics is becoming increasingly 'digital', as witnessed, for example, by the expanding use of pulse-code modulation for communications systems; and the digital computer is now widely used for the storage, analysis and processing of signals and data of all types. A further important reason for this emphasis lies in my belief that many central concepts of signal processing—for example convolution and correlation—are

more easily grasped when applied to sampled-data signals. This approach
has given me an excuse to include a substantial section on digital filters, which
makes this book one of the first to deal with this important topic at an introductory
level.

In rather more detail, chapters 2 and 3 give fairly conventional introductions to
signal analysis by Fourier and Laplace methods, although limitations of space make
it impossible to do proper justice to these powerful techniques. Chapter 4 deals
with sampled-data signals and introduces the z-transform: from this point on, no
special distinction is made between continuous and sampled-data signals, and
later chapters illustrate concepts by reference to either or both types of function.
The reader is therefore encouraged to tackle chapter 4 before proceeding, even if
his main interest lies in continuous (analogue) signals and systems. The remainder
of the book requires little introduction, except to say that the length of a
chapter reflects only the amount of material I have thought appropriate to include
under a particular heading. Chapter lengths therefore vary widely, and the longer
ones are not intended to be read (even less digested) at one sitting. Indeed, it may
be helpful to point out that the only part of chapter 5 ('Random Signals') which
is important for an understanding of later chapters is section 5.4, which deals with
autocorrelation and power spectral density functions: section 7.4 may also be
omitted by the reader with little interest in random signals.

It is a pleasure to thank my colleague, G. A. L. Reed, for his many helpful
suggestions and comments on the manuscript; and to record my debt to Professor
B. McA. Sayers, of the Engineering in Medicine Laboratory in the Electrical
Engineering Department at Imperial College, London, who first introduced me to
many of the ideas covered in this book. Finally, my thanks are due to Miss V.
Mackintosh and Miss L. Jackson who have tackled the typing with great
competence and good humour.

PAUL A. LYNN

Preface to the Second Edition

The nine years since this book was first published have witnessed the remarkable growth of microelectronics. In the preface to the first edition, I noted that electronics was becoming increasingly digital, and the major impact of micro-electronics has of course been in this area. Powerful microprocessors and computers, together with the digital signal analysis and processing techniques they offer, are now available at modest cost to almost anyone who requires them. This has been reflected in the large number of texts on digital systems, filters and processing which have appeared in the last five years or so. However, many of these texts (and, to an even greater extent, research papers) remain more or less inaccessible to the non-specialist, including the engineer 'brought up on' analogue circuits, systems and techniques. For this reason I hope that this book, which covers both analogue and digital approaches, and makes no special distinction between them, will continue to prove useful.

In view of the rapid developments in digital electronics, this second edition concentrates on revising the material on digital signal processing. The sections on digital filters in chapter 9 have been rearranged and expanded, a clear distinction now being made between filters of the finite impulse response (FIR) and infinite impulse response (IIR) types. This is in line with most other recent textbooks on digital filtering. Chapter 10 has been largely rewritten and its new title reflects its broader scope. In particular, I have included material on optimum Wiener and Kalman filtering — topics which, once again, the digital computer has given considerable practical interest. I feel that this final chapter may now reasonably claim to cover, even if only at an introductory level, most of the available linear techniques for signal recovery, estimation and prediction.

I should like to thank Mrs Angela Tyler for all her help with the revised manuscript.

PAUL A. LYNN

PART I

SIGNAL ANALYSIS

1

Background

1.1 Historical developments

Until very recent times, advances in the art and science of signalling have been
almost exclusively stimulated by the need for military intelligence. The use of
smoke signals and of drums for conveying information in situations where a
human messenger would be too slow goes back beyond recorded history, and
both Greeks and Romans were practised in the use of light beacons in the pre-
Christian era. Late sixteenth century England employed a long chain of beacons
to warn of the approach of the Spanish Armada, and at about this time the term
'signal' came into general use to denote 'a sign or notice, perceptible by sight or
hearing, given especially for the purpose of conveying warning, direction or
information'. By 1806 it was possible for a semaphore signal to be sent from
Plymouth to London, and acknowledged, in three minutes. The invention of the
famous signalling code by Samuel Morse in 1852 and the development of the
electric telegraph greatly enhanced the speed and the reliability with which
messages could be conveyed, and led to the widespread use of signalling for
purposes other than warfare and defence.

In all these early methods of transmitting information, signal theory, as we
know it, played no part. The signals used were very simple ones, and the receiver
was almost always the unaided human eye or ear. This situation changed rapidly
with the invention of the telephone and especially of radio, when for the first
time signals were transmitted and received by complex electrical apparatus, the
performance of which could be subjected to mathematical analysis. The modification of electrical signals representing messages as they pass through electrical
networks has become one of the central interests of signal theory. Another major
interest is the investigation of the ways in which a signal may be degraded in
passing along a transmission path, and of the techniques which may be used to
improve the chances of its correct detection at the receiver. This type of problem
has received a great deal of attention ever since the early days of radar, and the
fact that the disturbances to which a signal is subject are generally of a random
type has assured a key role for statistical analysis in modern signal theory.

3

From this brief discussion, it will be clear that major developments in the theory of signal analysis and processing have accompanied the recent rapid growth in electrical and electronic methods of communication. While it is probable that signal theory will continue to make a major contribution to communication techniques, it is now clear that it also has great potential in other fields. There are, in other words, great advantages to be gained by broadening the definition of the word 'signal' to include almost any physical variable of interest, and to extend the techniques of signal analysis and processing to other areas of enquiry.

1.2   Signal types and signal sources

Conventionally, any physical variable which represents the message in a communication system is thought of as a signal; typical examples are the voltage waveform present at the output terminal of an amplifier, or the current delivered to the coil of a loudspeaker. In a complex system, the signal which bears the information may well take on a variety of forms, being represented at one point by a pressure, at another by a current, at a third by a light intensity, and so on. Signals may be either continuous or discrete. An example of the first type is the current in a loudspeaker coil, which varies continuously as a function of time and which may take on an infinite variety of values. On the other hand, when a lamp is used to transmit the Morse Code the signal is discrete in the sense that it can take on only two possible values, corresponding to the lamp's being switched on or off. These two types of signal are shown in figure 1.1.

Some other examples of time-varying functions which may be regarded as signals and which arise in quite different contexts, are illustrated in figure 1.2. The all-or-none firing patterns of nerve fibres which relay information to and from the brain form an interesting class of discrete signal, in which the essential interest is in the timing of successive identical discharges. A complex economic system may be considered a very fruitful source of signals, such as interest rates or the unit cost of a manufactured article. Some signals are discrete not in the sense described above, but because they only have values at certain instants. A good example would be the

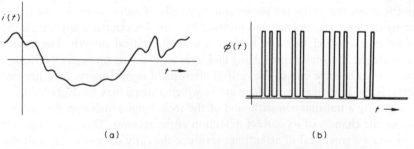

Figure 1.1   *Continuous and discrete signals. (a) represents the continuously varying current in a loudspeaker coil and (b) the light output from a lamp used to transmit a Morse Code message.*

midday temperature at a particular place measured on a number of successive days, although it is important to notice that in most cases such a signal may be considered as a sampled version of an underlying continuous function. Finally, it is perhaps useful to note that although most of the variables which may usefully be thought of as signals will be functions of time, there is no reason why they should be. There are certain occasions when a function such as that which describes, say, the variation of air temperature or humidity with altitude may usefully be thought

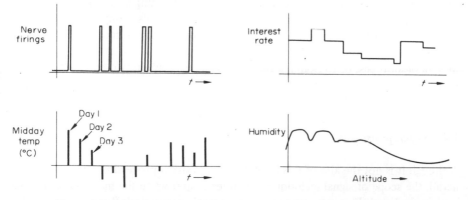

Figure 1.2   *Four functions which might usefully be considered as 'signals'.*

of as a signal. Indeed it is often best to think of a signal as a curve drawn on a piece of graph paper which has features of potential interest; that curve may be subjected to the techniques of signal analysis and processing, regardless of what in detail it represents.

1.3   The uses of signal theory

The value of using the concepts of signal analysis and processing in situations outside, as well as within, the field of communications may best be illustrated by considering the types of problem to which they are conventionally applied.

Signals are analysed for various reasons. It is, for example, sometimes possible to represent an apparently complex signal waveform by a limited set of parameters which, although not necessarily describing that waveform completely, are sufficient for the task in hand (such as deciding whether or not the signal may be faithfully transmitted through a particular communication channel). On other occasions, such a representation may be no more than a convenient shorthand method of signal description. But perhaps most important of all, careful analysis of a signal may often be used to learn something about the source which produced it; in other words, certain detailed characteristics of a signal which are not immediately apparent can often give important clues to the nature of the signal source, or to the type of processing which has occurred between that source and the point at which the signal is recorded or detected. This aspect of signal analysis is of obvious relevance to a great variety of situations in which causal relationships between a complex system and one of its measured 'output' variables are being investigated.

The last point leads naturally to the uses of signal processing. When the characteristics of a signal have been adequately defined, it is possible to determine the exact type of processing required to achieve a particular object. For example, it might be required to pass the signal undistorted through a communications system, to detect the occurrence of a particular signal waveform in the presence of random disturbances, or to extract by suitable processing some significant aspect of a signal or of the relationship between two signals. Once again, the techniques used to process signals in such ways are of interest in other fields; for example, it is often important to be able to clarify particular features or trends in data of all types, or to examine the relationships between two recorded variables.

1.4   Signal processors

Although it is not the aim of this book to discuss signal processing devices in any detail, the scope of signal techniques becomes clearer when the main classes of such devices are listed. By far the most important of these classes is the electronic circuit or system which is designed to analyse or to process signals of an electrical kind; examples of this type of device are legion, and include the radio or television set, the radar receiver, the laboratory wave analyser and the multiplexing equipment used in telephone exchanges. Very often, a physical quantity such as temperature or acceleration may be converted into an equivalent electrical signal by using a suitable transducer, and therefore processed or analysed using electronic circuits. The great versatility, accuracy, economy and speed of operation of electronic circuits makes their use for such purposes extremely attractive, although it is important to realise that hydraulic and pneumatic devices are also used to some extent. Purely mechanical equipments (such as the Henrici-Conradi and Michelson–Stratton harmonic analysers invented in the 1890s) have been rendered almost entirely obsolete by developments in electronics.

The practical uses of signal theory have been greatly extended in the past ten or fifteen years by the enormous increase in digital computing facilities available to most workers in applied science. It is now possible to store and process signals and data of all types using computer techniques, without the need for special-purpose electronic circuits, and indeed without the need for any understanding of electrical technology. As a result, almost any observable quantity, whether obtained from investigations in science or engineering, medicine, economics or the social sciences, may now be subjected to signal analysis procedures. This fact, more than any other, is responsible for the increasing interest of signal techniques in areas other than electronics and communications.

2

Periodic Signals

2.1 Time-domain descriptions

The fact that the great majority of functions which may usefully be considered as signals are functions of time lends justification to the treatment of signal theory in terms of time and of frequency. A periodic signal will therefore be considered to be one which repeats itself exactly every T seconds, where T is called the period of the signal waveform; the theoretical treatment of periodic waveforms assumes that this exact repetition is extended throughout all time, both past and future. Portions of such continuous periodic waveforms are illustrated in figure 2.1. In practice, of course, signals do not repeat themselves indefinitely. Nevertheless, a waveform such as the output voltage of a mains rectifier prior to smoothing[8,16] does repeat itself very many times, and its analysis as a strictly periodic signal yields valuable results. In other cases, such as the electrocardiogram, the waveform is quasi-periodic and may usefully be treated as truly periodic for some purposes. It

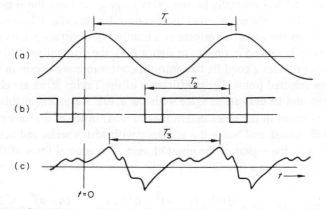

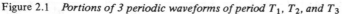

Figure 2.1   *Portions of 3 periodic waveforms of period T_1, T_2, and T_3*

is worth noting that a truly repetitive signal is of very little interest in a communications channel, since no further information is conveyed after the first cycle of the waveform has been received. One of the main reasons for discussing periodic signals is that a clear understanding of their analysis is a great help when dealing with aperiodic and random ones.

A complete time-domain description of such a signal involves specifying its value precisely at every instant of time. In some cases this may be done very simply using mathematical notation; for example, waveform (a) of figure 2.1 is completely specified by the function

$$f(t) = A \sin(\omega t + \phi)$$

Waveform (b) is also quite simple to express mathematically, whereas (c) is obviously highly complex.

Fortunately, it is in many cases useful to describe only certain aspects of a signal waveform, or to represent it by a mathematical formula which is only approximate. The following aspects might be relevant in particular cases:

(i) the average value of the signal,
(ii) the peak value reached by the signal,
(iii) the proportion of the total time spent between values y_1 and y_2,
(iv) the period of the signal.

If it is desired to approximate the waveform by a mathematical expression, such techniques as a polynomial expansion, a Taylor series, or a Fourier series may be used. A polynomial of order n having the form

$$f(t) = a_0 + a_1 t + a_2 t^2 + a_3 t^3 + \ldots a_n t^n$$

may be used to fit the actual curve at $(n + 1)$ arbitrary points, as shown in figure 2.2. The accuracy of fit will generally improve as the number of polynomial terms increases. It should also be noted that the error between the true signal waveform and the polynomial will normally become very large away from the region of the fitted points, and that the polynomial itself cannot be periodic. Whereas a polynomial approximation fits the actual waveform at a number of arbitrary points (which need not be equally spaced in time as in figure 2.2), the alternative Taylor series approximation provides a good fit to a smooth continuous waveform in the vicinity of one selected point. The coefficients of the Taylor series are chosen to make the series and its derivatives agree with the actual waveform at this point. The number of terms in the series determines to what order of derivative this agreement will extend, and hence the accuracy with which series and actual waveform agree in the region of the point chosen. The general form of the Taylor series for approximating a function $f(t)$ in the region of the point $t = a$ is given by

$$f(t) \simeq f(a) + (t - a) \times \frac{\mathrm{d}f(a)}{\mathrm{d}t} + \frac{(t - a)^2}{2!} \times \frac{\mathrm{d}^2 f(a)}{\mathrm{d}t^2} + \ldots \frac{(t - a)^n}{n!} \times \frac{\mathrm{d}^n f(a)}{\mathrm{d}t^n}$$

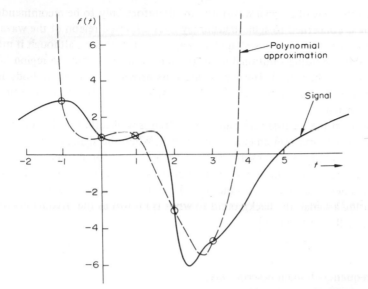

Figure 2.2 *Approximation of a signal waveform by a polynomial. Here the function $f(t) = (1 + t + 0.5\ t^2 - 2t^3 + 0.5\ t^4)$ has been fitted to a signal waveform at five points*

A simple example of the use of the series is illustrated in figure 2.3, in which the sinusoidal wave (sin t) is approximated in the region of $t = \pi/6$ by just the first three terms of the expansion. If the period of the wave is chosen as 1 second, the first three terms become

$$[0.5 + 5.44(t - 0.0833) - 19.7(t - 0.0833)^2]$$

As would be expected, the fit to the actual waveform is good in the region of the point chosen, but rapidly deteriorates to either side. The polynomial and Taylor

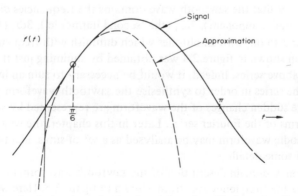

Figure 2.3 *Approximation of a signal waveform by a Taylor series. Here the first three terms of the series have been used to approximate the function (sin t) in the region t = π/6*

series descriptions of a signal waveform are therefore only to be recommended when one is concerned to achieve accuracy over a limited region of the waveform. The accuracy usually decreases rapidly away from this region, although it may be improved by including additional terms (so long as t lies within the region of convergence of the series). The approximations provided by such methods are never periodic in form and cannot therefore be considered ideal for the description of repetitive signals.

By contrast the Fourier series approximation is well suited to the representation of a signal waveform over an extended interval. When the signal is periodic, the accuracy of the Fourier series description is maintained for all time, since the signal is represented as the sum of a number of sinusoidal functions which are themselves periodic. Before examining in detail the Fourier series method of representing a signal, the background to what is known as the 'frequency-domain' approach will be introduced.

2.2   Frequency-domain descriptions

The basic concept of frequency-domain analysis is that a waveform of any complexity may be considered as the sum of a number of sinusoidal waveforms of suitable amplitude, periodicity, and relative phase. A continuous sinusoidal function (sin ωt) is thought of as a 'single frequency' wave of frequency ω radians/second, and the frequency-domain description of a signal involves its breakdown into a number of such basic functions. This is the method of Fourier analysis.

The way in which a periodic wave may be built up from a number of sinusoidal waves is illustrated in figure 2.4. The periodic waveform chosen is of 'sawtooth' form, and Fourier analysis shows that it may be represented by the summation of an infinite number of sinusoidal waves as follows

$$f(t) = \sin \omega_1 t - \tfrac{1}{2} \sin 2\omega_1 t + \tfrac{1}{3} \sin 3\omega_1 t - \tfrac{1}{4} \sin 4\omega_1 t + \dots$$

We therefore say that the sawtooth wave contains the frequencies ω_1 (known as the 'fundamental' component), $2\omega_1$ (the 'second harmonic'), $3\omega_1$ (the 'third harmonic'), and so on, with amplitudes which diminish with frequency. The approximation shown in figure 2.4 was obtained by summing just the first 4 terms of the above series. Indeed, it would be necessary to sum an infinite number of terms of the series in order to synthesise the sawtooth waveform perfectly; in particular, the sudden changes of the waveform are represented by very high frequency terms of the Fourier series. Later in this chapter, the way in which a complex periodic waveform may be analysed as a set of sinusoidal functions will be covered in some detail.

The frequency-domain description of the sawtooth waveform may be represented graphically by the 'frequency spectrum' shown in figure 2.5. Here we see at a glance that the wave may be considered to be made up from a sinusoidal wave of frequency ω_1 and amplitude 1·0 added to another sinusoidal wave of frequency

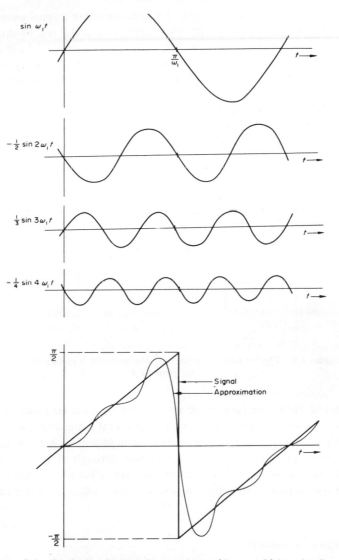

Figure 2.4 *Synthesis of a periodic waveform of 'sawtooth' form by the addition of a number of sinusoidal functions*

$2\omega_1$ and amplitude 0·5, and so on. In the particular case of the sawtooth waveform of figure 2.4, the various frequency components have very simple phase relationships, successive terms being inverted; in the general case, however, the phase relationships will be complicated and must be indicated—normally by drawing one graph to indicate the amplitudes of the various components, and another to indicate their relative phases.

There are a number of reasons why signal representation in terms of a set of component sinusoidal waves occupies such a central role in signal analysis. The suitability of a set of periodic functions for approximating a signal waveform over an extended interval has already been mentioned, and it will be shown later that the use of such techniques causes the error between the actual signal and its approximation to be minimised in a certain important sense. A further reason why sinusoidal functions are so important in signal analysis is that they occur widely in the physical world (simple harmonic motion, vibrating strings and structures, wave motion, and alternating electrical current) and are very susceptible to mathematical treatment; a large and extremely important class of electrical and mechanical systems, known as 'linear systems', responds sinusoidally when driven

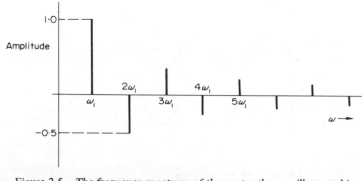

Figure 2.5   *The frequency spectrum of the sawtooth wave illustrated in figure 2.4*

by a sinusoidal disturbing function of any frequency. All these manifestations of sinusoidal functions in the physical world suggest that signal analysis in sinusoidal terms will simplify the problem of relating a signal to underlying physical causes, or to the physical properties of a system or device through which it has passed. Finally, sinusoidal functions form a set of what are called 'orthogonal functions', the rather special properties and advantages of which will now be discussed.

2.3 Orthogonal functions

2.3.1 Vectors and signals

A discussion of orthogonal functions and of their value for the description of signals may be conveniently introduced by considering the analogy between signals and vectors. A vector is specified both by its magnitude and direction, familiar examples being force and velocity. Suppose we have two vectors V_1 and V_2; geometrically, we define the component of vector V_1 along vector V_2 by constructing the perpendicular from the end of V_1 onto V_2 as shown in figure 2.6. We then have

$$V_1 = C_{12}V_2 + V_e$$

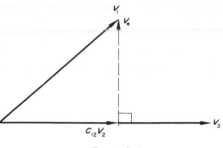

Figure 2.6

If we wish to approximate the vector V_1 by a vector in the direction of V_2, then the error in the approximation is the vector V_e. Clearly, this error vector is of minimum length when it is drawn perpendicular to the direction of V_2. Thus we say that the component of vector V_1 along vector V_2 is given by $C_{12}V_2$, where C_{12} is chosen such as to make the error vector as small as possible. If C_{12} is zero, then one vector has no component along the other, and the vectors are said to be orthogonal. Conversely if V_e is zero and $C_{12} = 1$, the two vectors are identical in both magnitude and direction. A familiar case of an orthogonal vector system is the use of three mutually-perpendicular axes in co-ordinate geometry.

These basic ideas about the comparison of vectors may be extended to signals. Suppose we wish to approximate a signal $f_1(t)$ by another signal or function $f_2(t)$ over a certain interval $t_1 < t < t_2$; in other words

$$f_1(t) \approx C_{12} . f_2(t) \text{ for } t_1 < t < t_2.$$

We wish to choose C_{12} to achieve the best approximation. If we define the error function

$$f_e(t) = f_1(t) - C_{12}f_2(t)$$

it might appear at first sight that we should choose C_{12} so as to minimise the average value of $f_e(t)$ over the chosen interval. The disadvantage of such an error criterion is that large positive and negative errors occurring at different instants would tend to cancel each other out. This difficulty is avoided if we choose to minimise the average squared-error, rather than the error itself (this is equivalent to minimising the square root of the mean-squared error, or 'r.m.s.' error). Denoting the average of $f_e^2(t)$ by ϵ, we have

$$\epsilon = \frac{1}{(t_2 - t_1)} \int_{t_1}^{t_2} f_e^2(t) \, dt = \frac{1}{(t_2 - t_1)} \int_{t_1}^{t_2} [f_1(t) - C_{12}f_2(t)]^2 \, dt$$

Differentiating with respect to C_{12} and putting the resulting expression equal to zero gives the value of C_{12} for which ϵ is a minimum. Thus

$$\frac{d}{dC_{12}} \left\{ \frac{1}{(t_2 - t_1)} \int_{t_1}^{t_2} [f_1(t) - C_{12} . f_2(t)]^2 \, dt \right\} = 0$$

Expanding the bracket and changing the order of integration and differentiation gives

$$\frac{1}{(t_2 - t_1)} \left[\int_{t_1}^{t_2} \frac{d}{dC_{12}} f_1^2(t) . dt - 2 \int_{t_1}^{t_2} f_1(t) . f_2(t) . dt + 2C_{12} \int_{t_1}^{t_2} f_2^2(t) . dt \right] = 0$$

The first integral vanishes since $f_1(t)$ is not a function of C_{12}, and hence

$$C_{12} = \frac{\displaystyle\int_{t_1}^{t_2} f_1(t) . f_2(t) . dt}{\displaystyle\int_{t_1}^{t_2} f_2^2(t) . dt}$$

By direct analogy with vectors, if C_{12} is zero we say that signal $f_1(t)$ contains no component of $f_2(t)$ and that the signals are therefore orthogonal over the interval $t_1 < t < t_2$. In this case it is clear that

$$\int_{t_1}^{t_2} f_1(t) . f_2(t) . dt = 0$$

Conversely if $f_1(t)$ and $f_2(t)$ are identical waveforms in the specified interval, C_{12} will equal unity.

As an example, suppose we wish to approximate the square wave shown in figure 2.7 by a sinusoidal wave having the same period, in the interval $0 < t < 2\pi/\omega$.

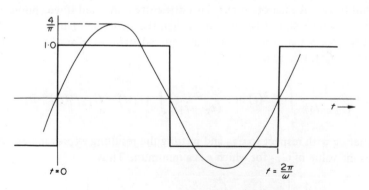

Figure 2.7   *A square wave and its approximation by a sine wave of the same period*

The square wave is defined in the interval $0 < t < 2\pi/\omega$ as

$$f_1(t) = 1, \qquad 0 < t < \frac{\pi}{\omega}$$

$$f_1(t) = -1, \qquad \frac{\pi}{\omega} < t < \frac{2\pi}{\omega}$$

The value of C_{12} which minimises the mean square error between the square wave and its approximation is therefore

$$C_{12} = \frac{\displaystyle\int_0^{2\pi/\omega} f_1(t) \cdot \sin \omega t \, dt}{\displaystyle\int_0^{2\pi/\omega} \sin^2 \omega t \cdot dt} = \frac{\displaystyle\int_0^{\pi/\omega} \sin \omega t \cdot dt + \int_{\pi/\omega}^{2\pi/\omega} (-\sin \omega t) \, dt}{\pi/\omega} = \frac{4}{\pi}$$

In other words $f_1(t) \approx (4/\pi) \sin \omega t$ is the approximation of the square wave by the sine wave which causes the least mean square error. The function $(4/\pi) \sin \omega t$ is shown superimposed upon the square wave in figure 2.7.

2.3.2 Signal description by sets of orthogonal functions

Suppose that we have approximated a signal $f_1(t)$ over a certain interval by the function $f_2(t)$ so that the mean square error is minimised, but that we now wish to improve the approximation. It will be demonstrated that a very attractive approach is to express the signal in terms of a set of functions $f_2(t), f_3(t), f_4(t)$, etc., which are mutually orthogonal. Suppose the initial approximation is

$$f_1(t) \approx C_{12} \cdot f_2(t)$$

and that the error is further reduced by putting

$$f_1(t) \approx C_{12} \cdot f_2(t) + C_{13} \cdot f_3(t)$$

where $f_2(t)$ and $f_3(t)$ are orthogonal over the interval of interest. Now that we have incorporated the additional term $C_{13} \cdot f_3(t)$, it is interesting to find what the new value of C_{12} must be in order that the mean square error is again minimised. We now have

$$f_e(t) = f_1(t) - C_{12} \cdot f_2(t) - C_{13} \cdot f_3(t)$$

and the mean square error in the interval $t_1 < t < t_2$ is therefore

$$\epsilon = \frac{1}{(t_2 - t_1)} \int_{t_1}^{t_2} [f_1(t) - C_{12} \cdot f_2(t) - C_{13} \cdot f_3(t)]^2 \cdot dt$$

Differentiating partially with respect to C_{12} to find the value of C_{12} for which the mean square error is again minimised, and changing the order of differentiation and integration, we have

$$\frac{\partial \epsilon}{\partial C_{12}} = \frac{1}{(t_2 - t_1)} \left[\int_{t_1}^{t_2} \frac{\partial}{\partial C_{12}} . f_1^2(t) . dt + \int_{t_1}^{t_2} 2C_{12} . f_2^2(t) . dt \right.$$

$$+ \int_{t_1}^{t_2} \frac{\partial}{\partial C_{12}} . C_{13}^2 . f_3^2(t) . dt - \int_{t_1}^{t_2} 2f_1(t) . f_2(t) . dt$$

$$\left. - \int_{t_1}^{t_2} \frac{\partial}{\partial C_{12}} . 2f_1(t) . C_{13} . f_3(t) . dt + \int_{t_1}^{t_2} 2f_2(t) . C_{13} . f_3(t) . dt \right]$$

The first, third and fifth integration terms are clearly zero, and the sixth one is also zero because $f_2(t)$ and $f_3(t)$ have been assumed orthogonal in the interval t_1 to t_2. Therefore, putting the whole expression to zero gives

$$C_{12} = \frac{\int_{t_1}^{t_2} f_1(t) . f_2(t) . dt}{\int_{t_1}^{t_2} f_2^2(t) . dt}$$

which is the same result as before. In other words, the decision to improve the approximation by incorporating an additional term in $f_3(t)$ does not require us to modify the coefficient C_{12}, provided that $f_3(t)$ is orthogonal to $f_2(t)$ in the chosen time interval. By precisely similar arguments we could show that the value of C_{13} would be unchanged if the signal were to be approximated by $f_3(t)$ alone.

This important result may be extended to cover the representation of a signal in terms of a whole set of orthogonal functions. The value of any coefficient does not depend upon how many functions from the complete set are used in the approximation, and is thus unaltered when further terms are included. The use of a set of orthogonal functions for signal description is analogous to the use of three mutually-perpendicular (that is, orthogonal) axes for the description of a vector in three-dimensional space, and gives rise to the notion of a 'signal space'. Accurate signal representation will often require the use of many more than three orthogonal functions, so that we must think of a signal within some interval $t_1 < t < t_2$ as being represented by a point in a multidimensional space.

It has already been stated at the end of section 2.2 that sinusoidal waves of different frequencies form an orthogonal set. In detail it may be shown[1] that a composite set of functions

$$\left.\begin{array}{l} \sin n\omega_1 t \\ \cos n\omega_1 t \end{array}\right\}, \qquad n = 1, 2, 3, 4 \ldots$$

and

are orthogonal in any interval $t_1 < t < (t_1 + (2\pi/\omega_1))$, that is, in any interval equal to one period of the lowest frequency wave $\sin \omega_1 t$ (or $\cos \omega_1 t$). We should also note that when $n = 0$ $\sin n\omega_1 t = 0$ and $\cos n\omega_1 t = 1$, and hence the complete orthogonal set comprises the functions

$$1, \cos \omega_1 t, \cos 2\omega_1 t \ldots \sin \omega_1 t, \sin 2\omega_1 t, \ldots \text{etc.}$$

Although the sinusoidal set of orthogonal functions is of the greatest practical value, it is certainly not the only one[18]. For example, the so-called Legendre polynomials form a set of mutually orthogonal functions over the interval $-1 < t < 1$, and are defined by the formula

$$P_n(t) = \frac{1}{2^n n!} \frac{d^n}{dt^n} (t^2 - 1)^n, \qquad n = 0, 1, 2, 3 \ldots$$

from which it follows that

$$P_0(t) = 1; \qquad P_1(t) = t; \qquad P_2(t) = (\tfrac{3}{2}t^2 - \tfrac{1}{2}); \qquad P_3(t) = (\tfrac{5}{2}t^3 - \tfrac{3}{2}t) \text{ etc.}$$

The orthogonality of this set distinguishes it from the simple power series in t used to fit an nth-order polynomial to a waveform at $(n + 1)$ points, as discussed previously. As with the sinusoidal set of functions, the use of the Legendre set gives rise to a minimum average squared-error criterion over the specified interval. Other orthogonal sets satisfy a 'weighted' squared-error criterion, in which the significance of the error between signal and its approximation depends upon where in the time interval that error occurs. For example, the Laguerre set of polynomials assumes the significance of the error to decrease exponentially in the range $0 < t <$ and minimises its mean squared value on this basis. By contrast, the Chebychev polynomial set, orthogonal in the interval $-1 < t < 1$, attaches relative importance to approximation errors occurring close to $t = \pm 1$.

Walsh functions form another orthogonal set, the importance of which is now being recognised for the description of signals and waveforms which take on a limited number (and often only two) distinct levels. Such waveforms are of increasing importance in digital electronic systems, including digital computers and other switching and logic circuits. The family of Walsh functions, originally described in 1923[19], are orthogonal in the range $0 < t < 1$, and each function is constant at a value ± 1 over each of a finite number of equal subintervals into which the interval $0 < t < 1$ is divided. A few of these functions are illustrated in figure 2.8. An nth-order Walsh function has $(n - 1)$ transitions between levels 1 and -1 in the specified interval, and as n increases, there exist more and more waveforms having this number of transitions which meet the criterion of orthogonality. Although it is beyond the scope of this book to discuss Walsh functions in any detail, it requires little imagination to appreciate their value for the type of application mentioned above.

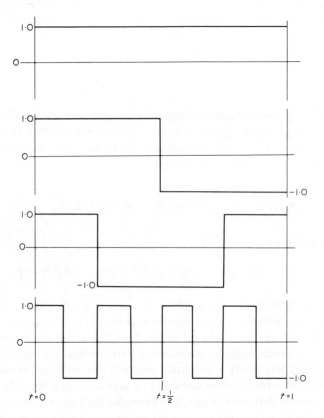

Figure 2.8   *Four examples of Walsh functions, which form an orthogonal set in the range $0 < t < 1$*

To summarise, there are a number of sets of orthogonal functions available for the approximate description of a signal waveform, of which the sinusoidal set is the most widely used. Sets involving polynomials in t are not by their very nature periodic, but may sensibly be used to describe one cycle (or less) of a periodic waveform; outside the chosen interval, errors between the true signal and its approximation will normally increase rapidly. A description of one cycle of a periodic signal in terms of sinusoidal functions will, however, be equally valid for all time because of the periodic nature of every member of the orthogonal set.

2.4 The Fourier series

Jean Baptiste Joseph, Baron de Fourier, was thirty years old when he accompanied
Napoleon on the latter's Egyptian campaign in 1798. He became Governor of
Lower Egypt and contributed many scientific papers to the Egyptian Institute
which Napoleon founded, but when the French army capitulated in 1801 he
returned to France and was made Prefect of the Department of Grenoble. He
continued his scientific researches and in 1822 published his monumental work
Théorie analytique de ia chaleur, in which he evolved the series which bears his
name. Originally applied to the analysis of heat flow, the series has since been
used in many branches of applied science, and constitutes one of the principal
tools of signal analysis.

The basis of the Fourier series is that a complex periodic waveform may be
analysed into a number of harmonically-related sinusoidal waves which constitute
an orthogonal set. If we have a periodic signal $f(t)$ with a period equal to T, then
$f(t)$ may be represented by the series

$$f(t) = A_0 + \sum_{n=1}^{\infty} A_n \cos n\omega_1 t + \sum_{n=1}^{\infty} B_n \sin n\omega_1 t$$

where $\omega_1 = 2\pi/T$. Thus $f(t)$ is considered to be made up by the addition of a
steady level (A_0) to a number of sinusoidal and cosinusoidal waves of different
frequencies. The lowest of these frequencies is ω_1 (radians per second) and is
called the 'fundamental'; waves of this frequency have a period equal to that of
the signal $f(t)$. Frequency $2\omega_1$ is called the 'second harmonic', $3\omega_1$ is the 'third
harmonic', and so on. Certain restrictions, known as the Dirichlet conditions[16],
must be placed upon $f(t)$ for the above series to be valid. The integral $\int |f(t)|\, dt$
over a complete period must be finite, and $f(t)$ may not have more than a finite
number of discontinuities in any finite interval. Fortunately, these conditions
do not exclude any signal waveform of practical interest.

2.4.1 Evaluation of the coefficients

We now turn to the question of evaluating the coefficients A_0, A_n and B_n. Using
the minimum square error criterion described in section 2.3.1, and writing $(\omega_1 t)$
as x for the sake of convenience, we have

$$A_0 = \frac{\int_{-\pi}^{\pi} f(x) . 1 . dx}{\int_{-\pi}^{\pi} 1 . dx} = \frac{1}{2\pi} . \int_{-\pi}^{\pi} f(x) . dx$$

$$A_n = \frac{\int_{-\pi}^{\pi} f(x) \cdot \cos nx \cdot dx}{\int_{-\pi}^{\pi} \cos^2 nx \cdot dx} = \frac{1}{\pi} \int_{-\pi}^{\pi} f(x) \cos nx \cdot dx$$

$$B_n = \frac{\int_{-\pi}^{\pi} f(x) \cdot \sin nx \cdot dx}{\int_{-\pi}^{\pi} \sin^2 nx \cdot dx} = \frac{1}{\pi} \cdot \int_{-\pi}^{\pi} f(x) \cdot \sin nx \cdot dx$$

Although in the majority of cases it is convenient for the interval of integration to be symmetrical about the origin, any interval equal in length to one period of the signal waveform may be chosen.

Many waveforms of practical interest are either even or odd functions of time. If $f(t)$ is even then by definition $f(t) = f(-t)$, whereas if it is odd $f(t) = -f(-t)$. If $f(t)$ is even and we multiply it by the odd function $\sin n\omega_1 t$ the result is also odd. Thus the integrand for every B_n is odd. Now when an odd function is integrated over an interval symmetrical about $t = 0$, the result is always zero. Hence all the B coefficients are zero and we are left with a series containing only cosines. By similar arguments, if $f(t)$ is odd the A coefficients must be zero and we are left with a sine series. It is indeed intuitively clear that an even function can only be built up from a number of other functions which are themselves even, and vice versa. As an example, we now evaluate the coefficients of the sawtooth wave already illustrated in figure 2.4, which is odd. In this case, all A_n will be zero, except possibly A_0. The waveform is given by $f(t) = \omega_1 t/2$ in the interval $-\pi/\omega_1 < t < \pi/\omega_1$; replacing $\omega_1 t$ by x for convenience, and changing the limits to $x = \pm\pi$, we have

$$A_0 = \frac{1}{2\pi} \int_{-\pi}^{\pi} f(x) \cdot dx = \frac{1}{2\pi} \int_{-\pi}^{\pi} \frac{x}{2} dx = \frac{1}{8\pi} [x^2]_{-\pi}^{\pi} = 0$$

and

$$B_n = \frac{1}{\pi} \int_{-\pi}^{\pi} f(x) \cdot \sin nx \cdot dx = \frac{1}{\pi} \int_{-\pi}^{\pi} \frac{x}{2} \cdot \sin nx \cdot dx$$

which may be integrated by parts to give

$$\frac{1}{2\pi} \left[\frac{\sin nx}{n^2} - \frac{x \cos nx}{n} \right]_{-\pi}^{\pi} = \frac{1}{\pi n^2} (\sin n\pi - n\pi \cos n\pi)$$

If n is an odd integer, $\sin n\pi = 0$ and $\cos n\pi = -1$, giving $B_n = 1/n$; if n is an even integer, $\sin n\pi = 0$ and $\cos n\pi = 1$, giving $B_n = -1/n$. Thus

$$B_1 = 1, \qquad B_2 = -\tfrac{1}{2}, \qquad B_3 = \tfrac{1}{3}, \qquad B_4 = -\tfrac{1}{4}, \ldots$$

which gives the expansion of the sawtooth wave as a Fourier series which has already been quoted in section 2.2

$$f(t) = \sin \omega_1 t - \tfrac{1}{2} \sin 2\omega_1 t + \tfrac{1}{3} \sin 3\omega_1 t - \tfrac{1}{4} \sin 4\omega_1 t + \ldots.$$

A_0 is zero because the integral of the waveform over one complete period is zero— in other words it has zero average, or mean, value. In electrical engineering language such a waveform is said to possess no d.c. (that is, direct current) component.

We have already seen how the Fourier series is simplified in the case of an even or odd function, by losing either its sine or its cosine terms. A different type of simplification occurs in the case of a waveform possessing what is known as 'half-wave symmetry'. A number of waveforms are illustrated in figure 2.9, and all but two of them exhibit such symmetry. In mathematical terms, half-wave symmetry exists when

$$f(t) = -f(t + T/2).$$

In other words any two values of the waveform separated by $T/2$ will be equal in magnitude and opposite in sign. Figure 2.9 shows that a sine wave of period T, and its third harmonic, both have this property, whereas its second harmonic does not. Generalising, only odd harmonics exhibit half-wave symmetry, and therefore a waveform of any complexity which has such symmetry cannot contain even harmonic components. Conversely, a waveform known to contain any second, fourth, or other even harmonic components cannot display half-wave symmetry.

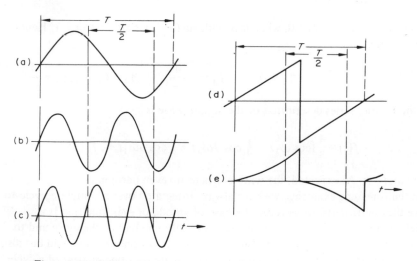

Figure 2.9 *Half-wave symmetry. Waveforms (a), (c) and (e) exhibit such symmetry; waveforms (b) and (d) do not*

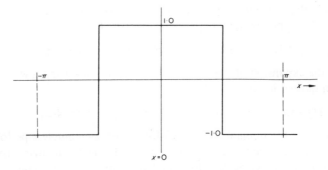

Figure 2.10   *A square waveform having half-wave symmetry*

To illustrate the absence of even-order harmonics in a wave with half-wave symmetry, we now evaluate the series coefficients for the square wave of figure 2.10. Its average value is clearly zero, giving $A_0 = 0$. Also all B_n are zero because the wave is an even function. The cosine coefficients are given by

$$A_n = \frac{1}{\pi} \int_{-\pi}^{\pi} f(x) . \cos nx . dx$$

$$= \frac{1}{\pi} \left\{ \int_{-\pi/2}^{\pi/2} \cos nx . dx + \int_{-\pi}^{-\pi/2} - \cos nx . dx + \int_{\pi/2}^{\pi} - \cos nx . dx \right\}$$

$$= \frac{4}{n\pi} \sin \left(\frac{n\pi}{2} \right)$$

When n is even $\sin(n\pi/2) = 0$; when n is odd, $\sin(n\pi/2)$ is either 1 or -1. Hence the coefficients are

$$A_0 = 0, \qquad A_1 = \frac{4}{\pi}, \qquad A_2 = 0, \qquad A_3 = -\frac{4}{3\pi}, \qquad A_4 = 0, \qquad A_5 = \frac{4}{5\pi} \dots$$

and the Fourier series description of the square wave is

$$f(t) = \frac{4}{\pi} \{\cos \omega_1 t - \tfrac{1}{3} \cos 3\omega_1 t + \tfrac{1}{5} \cos 5\omega_1 t + \dots\}$$

where $\omega_1 t = x$. As expected, the wave contains no even harmonics.

In the foregoing examples, we have always integrated over a complete cycle to derive the coefficients. However in the case of an odd or even function it is sufficient, and often simpler, to integrate over only one half of the cycle and to multiply the result by 2. Furthermore if the wave is not only even or odd but also displays halfwave symmetry, it is enough to integrate over one quarter of a cycle and multiply by 4. These closer limits are adequate in such cases because the

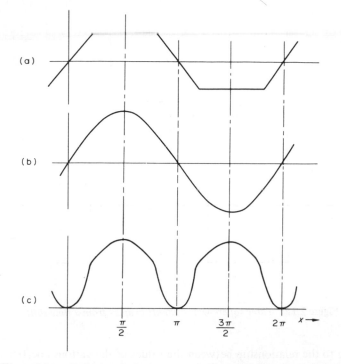

Figure 2.11   *Reducing the limits of integration. Waveform (a) is an odd function and displays half-wave symmetry. Multiplication by sine wave (b) gives the waveform in (c). When integrating (c) between 0 and 2π, it is sufficient to integrate between 0 and $\pi/2$ and multiply the result by 4*

function that is being integrated is repetitive, repeating twice within one period when the function is either even or odd, and four times within one period when it also exhibits half-wave symmetry; this point is illustrated in figure 2.11.

2.4.2 Choice of time origin, and waveform power

The amount of work involved in calculating the Fourier series coefficients for a particular waveform shape is reduced if the waveform is either even or odd, and this may often be arranged by a judicious choice of time origin. For example, figure 2.12 shows three versions of a square wave which differ only in their time origin. Wave (a) is an even function, symmetrical about $t = 0$. We have already seen in section 2.3.1 that its fundamental component is $(4/\pi) \cos \omega_1 t$. Wave (b) is identical except that it is an odd function with a fundamental equal to $(4/\pi) \sin \omega_1 t$. This shift of time origin has therefore merely had the effect of converting a Fourier series containing only sine terms into one containing only cosine terms; the amplitude of a component at any one frequency is, as we would expect, unaltered. The situation in (c) is however more complicated because the square wave is neither even nor odd, and must therefore be expected to include both sine and cosine terms in its Fourier series.

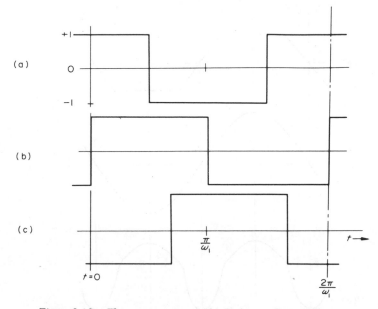

Figure 2.12   *Three square waves, identical apart from a time-shift*

The clue to the relationship between the values of the various coefficients in case (c) and those in (a) and (b) lies in the average power of the waveform, a concept familiar to electrical engineers. Suppose we find when we analyse the waveform of figure 2.12(c) that there are fundamental components

$$A_1 \cos \omega_1 t \qquad \text{and} \qquad B_1 \sin \omega_1 t$$

If the component $A_1 \cos \omega_1 t$ represents a voltage applied to a resistor of value 1 ohm, then the average power P dissipated by it in the resistor over one complete period will be:

$$P = \frac{1}{2\pi} \int\limits_{-\pi}^{\pi} (A_1 \cos x)^2 \, dx, \qquad \text{where} \qquad x = \omega_1 t$$

$$= \frac{A_1^2}{2}$$

In other words the mean power dissipated is equal to the average squared value of the voltage waveform[8]. Similarly, the average squared value of the wave $B_1 \sin \omega_1 t$ over one period is $B_1^2/2$. The total power represented by the two fundamental components together is thus $\frac{1}{2}(A_1^2 + B_1^2)$. It is clear however that this value will be the same for all three examples of the square wave of figure 2.12, since the average power represented by a waveform is not altered by a mere shift in time origin. Since for waveform (a) we have already found that $A_1 = 4/\pi$ and $B_1 = 0$, and for waveform (b) $B_1 = 4/\pi$ and $A_1 = 0$, we conclude that for any other waveform such as (c)

$$\left(\frac{A_1{}^2 + B_1{}^2}{2}\right) = \left(\frac{4}{\pi}\right)^2 \cdot \tfrac{1}{2}$$

$$\therefore \quad A_1{}^2 + B_1{}^2 = \left(\frac{4}{\pi}\right)^2$$

Hence as the time origin of a waveform is shifted, the various sine and cosine coefficients of its Fourier series will change, but the sum of the squares of any two coefficients A_n and B_n will remain constant.

The above ideas lead naturally to an alternative trigonometric form for the Fourier series. If the two fundamental components of a waveform are

$$A_1 \cos \omega_1 t \qquad \text{and} \qquad B_1 \sin \omega_1 t$$

their sum may be expressed in an alternative form using trigometric identities

$$A_1 \cos \omega_1 t + B_1 \sin \omega_1 t = \sqrt{(A_1{}^2 + B_1{}^2)} \cos \left(\omega_1 t - \tan^{-1} \frac{B_1}{A_1}\right)$$

$$= \sqrt{(A_1{}^2 + B_1{}^2)} \sin \left(\omega_1 t + \tan^{-1} \frac{A_1}{B_1}\right)$$

Thus the sine and cosine components at a particular frequency are expressed as a single cosine or sine wave together with a phase shift. This equivalence is illustrated in figure 2.13. If this procedure is applied to all harmonic components of the Fourier series, we get the alternative forms

$$f(t) = A_0 + \sum_{n=1}^{\infty} C_n \cos (n\omega_1 t - \phi_n)$$

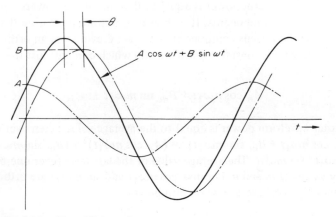

Figure 2.13   *The addition of a sine and a cosine function to give a wave of the same frequency with a phase angle θ*

or

$$f(t) = A_0 + \sum_{n=1}^{\infty} C_n \sin (n\omega_1 t + \theta_n)$$

where

$$C_n = \sqrt{(A_n^2 + B_n^2)}$$

$$\phi_n = \tan^{-1} \frac{B_n}{A_n}, \quad \text{and} \quad \theta_n = \tan^{-1} \frac{A_n}{B_n}$$

Finally, we note that since the mean power represented by any component wave is

$$\frac{A_n^2 + B_n^2}{2} = \frac{C_n^2}{2}$$

and the power represented by the term A_0 is simply A_0^2, the total average waveform power is equal to

$$P = A_0^2 + \tfrac{1}{2} \sum_{n=1}^{\infty} C_n^2$$

But P may also be expressed as the average value over one period of $[f(t)]^2$, using again the convention that $f(t)$ is considered to represent a voltage waveform applied across a 1 ohm resistor. Hence

$$P = A_0^2 + \tfrac{1}{2} \sum_{n=1}^{\infty} C_n^2 = \frac{1}{T} \int_{-T/2}^{T/2} [f(t)]^2 . dt$$

This result is a version of a more general one known as Parseval's theorem[2,4], and shows that the total waveform power is equal to the sum of the powers represented by its individual Fourier components. It is, however, important to note that this is only true because the various component waves are drawn from an orthogonal set. This may be shown by considering a wave $f(t)$ which contains just the two components

$$A_n \cos n\omega_1 t + B_m \sin m\omega_1 t$$

The average total waveform power is equal to the average value taken over one period of $(A_n \cos n\omega_1 t + B_m \sin m\omega_1 t)^2 = (A_n \cos n\omega_1 t)^2 + (B_m \sin m\omega_1 t)^2 + 2A_n B_m \cos n\omega_1 t . \sin m\omega_1 t$. The average value of the last term (over one period) is zero for any values of m and n, because $\cos n\omega_1 t$ and $\sin m\omega_1 t$ are orthogonal functions. Hence the total average waveform power is

$$\frac{A_n^2 + B_m^2}{2}$$

and is equal to the sum of the powers in the two individual components. A similar result is obtained for more complex waveforms having more Fourier components, since they are all members of an orthogonal set.

2.4.3 Some general comments

It is often possible to anticipate the main characteristics of the spectrum of a periodic waveform just from a visual inspection. For example, a waveform which has equal areas above and below the axis, such as that shown in figure 2.14(a), will have no zero-frequency component ($A_0 = 0$). A signal which exhibits sudden changes or discontinuities such as that of figure 2.14(b) must be expected to be rich in the higher-order harmonics, because it is only possible to build up such a

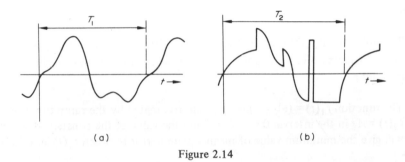

Figure 2.14

waveform using component waves which are themselves changing rapidly (that is, high frequency waves). This point is well illustrated by considering the square wave of figure 2.15(a), the Fourier series of which has already been found to be

$$f(t) = \frac{4}{\pi} (\cos \omega_1 t - \tfrac{1}{3} \cos 3\omega_1 t + \tfrac{1}{5} \cos 5\omega_1 t - \ldots)$$

If we integrate the square wave with respect to time (which is equivalent to summing the positive and negative areas under the waveform), we get the triangular

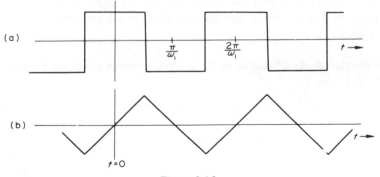

Figure 2.15

wave shown in figure 2.15(b). Term by term integration of the above Fourier series yields a new series which corresponds to the triangular wave

$$\frac{4}{\omega_1\pi}(\sin \omega_1 t - \tfrac{1}{9} \sin 3\omega_1 t + \tfrac{1}{25} \sin 5\omega_1 t - \ldots)$$

Thus the high-frequency components of the triangular wave are progressively smaller in magnitude than those of the square wave, a result which would be expected since the latter exhibits much sharper changes. A further integration would yield a still more rounded waveform containing high-frequency components of yet smaller amplitudes.

Problems

1. The polynomial $f_1(t) = a + bt + ct^2$ is to be fitted to the function $f_2(t) = e^t$ at the instants $t = 0$, $t = 1$, and $t = 2$. Find the required values for the constants a, b and c.

2. The function $f_1(t) = (e^t - 1)$ is to be approximated by the ramp function $f_2(t) = At$ in the interval $0 < t < 1$. Find the value of the constant A which will give the minimum value of mean squared-error between $f_1(t)$ and $f_2(t)$.

3. Prove that the two Legendre polynomials

$$P_1(t) = t \quad \text{and} \quad P_2(t) = (\tfrac{3}{2}t^2 - \tfrac{1}{2})$$

are orthogonal in the interval $-1 < t < 1$, and by sketching them satisfy yourself that this result is plausible.

4. The sawtooth waveform illustrated in figure 2.4 has the Fourier series

$$f(t) = \sin \omega_1 t - \tfrac{1}{2} \sin 2\omega_1 t + \tfrac{1}{3} \sin 3\omega_1 t - \tfrac{1}{4} \sin 4\omega_1 t + \ldots$$

Tabulate and plot the first four components of the series and add them, and hence show that it is likely that the infinite series will converge to the given waveform.

5. The diagram shows a function in the interval between $x = 0$ and $x = \pi/2$.

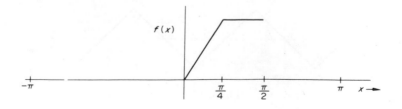

Complete the function over the full period from $x = -\pi$ to $x = \pi$ in such a way that:

   (i) the Fourier series contains only cosine terms,
   (ii) the Fourier series contains only sine terms,
   (iii) the function displays half-wave symmetry.

6. Determine from symmetry considerations what terms will be absent from the Fourier series of the function shown.

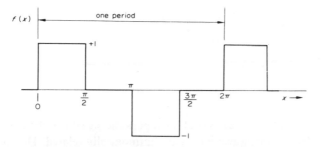

7. Find the Fourier series for the function $f(t) = (\omega t/\pi)^2$ in the interval $-\pi < \omega t < \pi$, given that the function repeats with a period of 2π.

8. Show that the Fourier series for the 'half-wave rectified' sine wave shown in the figure (that is, a sine wave with alternate half-cycles eliminated) is given by

$$f(t) = \frac{1}{\pi} + \frac{1}{2}\cos \omega_1 t + \frac{2}{3\pi}\cos 2\omega_1 t - \frac{2}{15\pi}\cos 4\omega_1 t + \dots$$
$$+ \frac{2(-1)^{1+n/2}}{\pi(n^2-1)}\cos n\omega_1 t + \dots (n \text{ even})$$

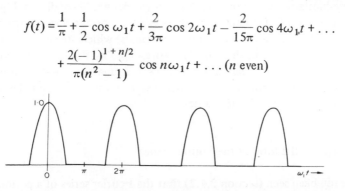

9. A wave $f_1(t) = 0.5 \sin \omega_1 t$ is added to another wave $f_2(t) = \cos \omega_1 t$. Express their sum in the form of a single waveform of sinusoidal form with an associated phase angle. Show that your result is reasonable by tabulating values of $f_1(t)$ and $f_2(t)$ and summing them graphically.

10. Show that approximately 90 per cent of the power (or energy) of a square wave having zero average value is contained in its first and third harmonic components.

3

Aperiodic Signals

3.1 Introduction

In the previous chapter we have seen how a periodic signal may be expressed as the sum of a set of sinusoidal waves which are harmonically related. The spectrum of such a signal consists of a number of discrete frequencies and is known as a 'line' spectrum. Although the analysis of periodic signals gives results which can be of great practical interest, the great majority of signals are not of this type. Firstly, even signals which repeat themselves a very large number of times are generally turned 'on' and 'off'. In other words they may not generally be assumed to exist for all time past, present, and future, and it is important to understand the effects which time-limitation has upon their frequency spectra. Secondly, and quite apart from any question of time-limitation, there is an important class of signal waveforms (amongst which are included random signals) which are simply not repetitive in nature and which cannot therefore be represented by Fourier series containing a number of harmonically-related frequencies. Fortunately, however, it is possible to derive frequency spectra for such signals using as a starting point the work we have already done on the Fourier series.

3.2 The exponential form of the Fourier series

It has already been seen (section 2.4.2) that the Fourier series of a periodic signal may be expressed in two ways: either as a set of sine and cosine waves of appropriate amplitude and frequency, or as a set of waves of sinusoidal form which are defined by their amplitudes and relative phase angles. A third form of the Fourier series, the exponential form, will now be discussed, since it is particularly helpful for the derivation of frequency spectra of aperiodic signals.

Instead of using the simple trigonometric form

$$f(t) = A_0 + \sum_{n=1}^{\infty} A_n \cos n\omega_1 t + \sum_{n=1}^{\infty} B_n \sin n\omega_1 t$$

we may write the exponential form

$$f(t) = \ldots + a_{-2}\exp(-j2\omega_1 t) + a_{-1}\exp(-j\omega_1 t) + a_0 + a_1\exp(j\omega_1$$

$$+ a_2\exp(2j\omega_1 t) + \ldots$$

$$= \sum_{m=-\infty}^{\infty} a_m\exp(jm\omega_1 t)$$

where m is any integer. Although these two forms look rather different, that they are in fact the same may be shown by using the identities

$$\cos x = \tfrac{1}{2}(e^{jx} + e^{-jx}) \text{ and } \sin x = \frac{-j}{2}(e^{jx} - e^{-jx})$$

Substitution into the first equation followed by rearrangement of terms yields the exponential form without difficulty. The coefficients of the two forms are related as follows

$$a_0 = A_0; a_m = \tfrac{1}{2}(A_m - jB_m), \qquad \text{when } m \text{ is positive}$$

and

$$a_m = \tfrac{1}{2}(A_m + jB_m), \qquad \text{when } m \text{ is negative.}$$

These results show that the coefficients of the exponential series are in general complex, and that they occur in conjugate pairs (that is, the imaginary part of a coefficient a_n is equal but opposite in sign to that of coefficient a_{-n}). Although the introduction of complex coefficients is at first difficult to understand, it should be remembered that the real part of a pair of coefficients denotes the magnitude of the cosine wave of the relevant frequency, and that the imaginary part denotes the magnitude of the sine wave. If a particular pair of coefficients a_n and a_{-n} are real, then the component at the frequency $n\omega_1$ is simply a cosine; if a_n and a_{-n} are purely imaginary, the component is just a sine; and if, as is the general case, a_n and a_{-n} are complex, both a cosine and a sine term are present.

The use of the exponential form of the Fourier series gives rise to a further notion which is often found difficult, that of 'negative frequency'. Of course, a cosine $A\cos\omega t$ is a wave of a single frequency ω radians/second, and may be represented by a single line of height A in a spectral diagram. If however we are using the exponential form of the Fourier series and are discussing a waveform in terms of its exponential components, we use the identity

$$A\cos\omega t = \frac{A}{2}(e^{j\omega t} + e^{-j\omega t})$$

Plotting the exponential components on a spectral diagram, we now consider the term $Ae^{j\omega t}/2$ to be represented by a line of height $A/2$ at a frequency ω, and the term $Ae^{-j\omega t}/2 = Ae^{j(-\omega)t}/2$ to be represented by a line of height $A/2$ at a frequency $-\omega$. Thus our frequency scale is now formally extended to include negative as well as positive frequencies, and a cosine component in a signal waveform gives rise to two spectral lines. Similarly, a sine component gives rise to two equal but opposite imaginary exponential components, which cannot of course be plotted on the same spectral diagram as the real exponential components representing the cosines. So a complete spectral description of a signal waveform will normally involve two separate diagrams, one representing real exponential terms (cosines) and the other representing imaginary terms (sines), as shown in figure 3.1. (Alternatively, the sine and cosine components may be combined to give a single amplitude term together with an associated phase angle, as already indicated in section 2.4.2.) It is therefore important to remember that the introduction of 'negative' frequencies implies that sines and cosines are being represented in exponential form.

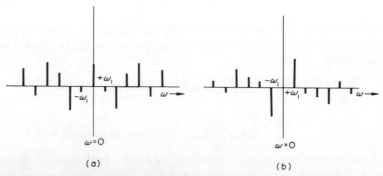

(a)   (b)

Figure 3.1 *Exponential representation of a Fourier Series; (a) Real parts of the exponential coefficients representing cosine components, and (b) imaginary parts representing sine components*

It has already been shown how the coefficients of the exponential Fourier series are related to the sine and cosine coefficients of the trigonometric form of the series, and although it is quite possible to derive them in this way it is not normally very convenient to do so. Recalling that the exponential coefficient a_m is given by $a_m = \frac{1}{2}(A_m - jB_m)$ for positive m, and that

$$A_m = \frac{1}{\pi} \int_{-\pi}^{\pi} f(x) \cos mx \, . \, dx$$

and $B_m = \frac{1}{\pi} \int_{-\pi}^{\pi} f(x) \sin mx \, . \, dx$, we may write

$$a_m = \frac{1}{2} \cdot \frac{1}{\pi} \left[\int_{-\pi}^{\pi} f(x) \cos mx \, . \, dx - j \int_{-\pi}^{\pi} f(x) \sin mx \, . \, dx \right]$$

$$= \frac{1}{2\pi} \int_{-\pi}^{\pi} f(x) (\cos mx - j \sin mx) \, . \, dx$$

$$= \frac{1}{2\pi} \int_{-\pi}^{\pi} f(x) \, e^{-jmx} \, dx, \qquad \text{where } x = \omega_1 t$$

When m is negative, $a_m = \frac{1}{2}(A_m + jB_m)$, and the above result also applies, as indeed it also does when $m = 0$ and the term e^{-jmx} becomes unity. Hence the result applies for all integer values of m, positive, negative, or zero, and shows how the exponential Fourier coefficients may be derived directly by multiplying the signal waveform by $e^{-jmx}/2\pi$ and integrating the result over a complete period. If the waveform exhibits half-wave symmetry (see section 2.4.1), the interval of integration may be shortened to half a period, just as it can when using the trigonometric form of the series. On the other hand, the interval of integration may not be shortened on account of the waveform being either even or odd; this was possible when using the trigonometric form because cosine and sine waves are odd and even respectively, but e^{-jmx} is neither even nor odd and such a simplification is not therefore admissible.

We now demonstrate the use of the exponential form of the Fourier series by two examples, the first of which is shown in figure 3.2(a). Since the triangular wave

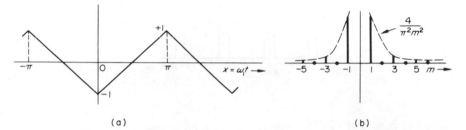

(a)   (b)

Figure 3.2   *(a) A triangular wave, and (b) the magnitudes of its real exponential Fourier coefficients*

exhibits half-wave symmetry, the coefficients a_m will be zero when m is even, and it is sufficient to integrate over the interval $x = 0$ to π. Over this interval, the wave is described by

$$f(x) = \left(\frac{2x}{\pi} - 1\right), \qquad 0 < x < \pi$$

hence

$$a_m = \frac{1}{\pi} \int_0^\pi \left(\frac{2x}{\pi} - 1\right) e^{-jmx} \, dx, \qquad \text{for odd } m.$$

Integration by parts gives

$$a_m = \frac{2}{\pi^2 m^2} (e^{jm\pi} - 1) + \frac{j}{m\pi} (e^{-jm\pi} + 1)$$

Now when m is odd, $e^{-jm\pi} = -1 = e^{jm\pi}$, and hence

$$a_m = -\frac{4}{\pi^2 m^2}, \qquad \text{for odd } m.$$

Conversely the original triangular waveform may be resynthesised by summing the following series

$$f(x) = -\frac{4}{\pi^2}\left(\ldots + \frac{e^{-j3x}}{9} + e^{-jx} + e^{jx} + \frac{e^{j3x}}{9} + \ldots\right)$$

which may be more conveniently written as

$$f(x) = -\frac{4}{\pi^2}\sum \frac{1}{m^2} \cdot e^{jmx}, \qquad \text{for odd } m.$$

We notice that the a_m coefficients have purely real values, which is expected since the wave is even. The spectrum may therefore be represented on a single diagram as shown in figure 3.2(b).

We now turn our attention to the analysis of the recurrent pulse waveform of figure 3.3(a). This wave is important for two main reasons: firstly, it is of great

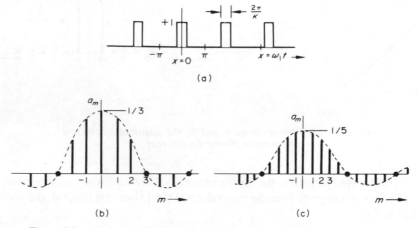

Figure 3.3   *(a) A repetitive pulse waveform, and its real exponential Fourier coefficients for (b) k = 3; and (c) k = 5*

practical interest because similar waveforms occur widely in such devices as digital computers and radar and communication systems; and secondly it is of analytical interest because it provides a good starting point for a discussion of the Fourier transform. For convenience we will assume (as in the diagram) that the period of the waveform is K times as large as the pulse duration. If we consider the period of the waveform between $x = \pm\pi$, it is clear that a finite contribution to the integral occurs only in the interval $x = \pm\pi/K$, where the pulse height is unity. Hence

$$a_m = \frac{1}{2\pi}\int_{-\pi/K}^{\pi/K} e^{-jmx} \cdot dx, \qquad \text{where } x = \omega_1 t$$

If $m = 0$, $\qquad a_0 = \dfrac{1}{2\pi}\left(\dfrac{\pi}{K} + \dfrac{\pi}{K}\right) = \dfrac{1}{K}$

If $m \neq 0$, $\qquad a_m = \dfrac{1}{-jm2\pi}\left[e^{-jmx}\right]_{-\pi/K}^{\pi/K} = \dfrac{1}{m\pi}\cdot\dfrac{e^{jm\pi/K} - e^{-jm\pi/K}}{2j}$

$$= \dfrac{1}{K}\dfrac{\sin(m\pi/K)}{m\pi/K}$$

Conversely, the recurrent pulse may be synthesised by summing components as follows

$$f(x) = \sum_{m=-\infty}^{m=\infty} \frac{1}{K}\cdot\frac{\sin(m\pi/K)}{m\pi/K}\cdot e^{jmx}$$

Figure 3.3(b) and (c) illustrates this result for $K = 3$ and $K = 5$. As K increases, the harmonic terms become closer spaced under the $(\sin x/x)$ 'envelope' and they reduce correspondingly in absolute size; the actual frequency represented by any line depends, of course, on the absolute period of the time function. If K becomes very large indeed, so that the pulse duration is very small compared with the waveform period, the spectrum consists of a correspondingly large number of spectral lines, very closely bunched and of vanishingly small amplitude. In the limit the lines become so close that we call the spectrum 'continuous', and we are led with little difficulty to the concepts of the Fourier transform.

3.3 The Fourier transform

3.3.1 Derivation

The Fourier transform (also called the Fourier integral) does for the non-repetitive signal waveform what the Fourier series does for the repetitive one. We have just seen how the line spectrum of a recurrent pulse waveform is modified as the pulse duration decreases, assuming the period of the waveform (and hence its fundamental component) remains unchanged. Suppose now that the duration of the pulses remains fixed but the separation between them increases, giving rise to an increasing period. In the limit, we will be left with a single rectangular pulse, its neighbours having moved away on either side towards $\pm\infty$. In this case the fundamental frequency ω_1 tends towards zero and the harmonics become extremely closely spaced and of vanishingly small amplitudes. Once again, we are left with a continuous spectrum.

Mathematically, this situation may be expressed by modifications to the exponential form of the Fourier series already derived

$$f(t) = \sum_{m=-\infty}^{\infty} a_m \exp(jm\omega_1 t)$$

where the complex coefficients a_m are found by multiplying $f(t)$ by $\exp(jm\omega_1 t)$ and taking the average of the result over a complete period

$$a_m = \frac{1}{2\pi} \int_{-\pi}^{\pi} f(x) . \exp(-jmx) . dx = \frac{1}{T} \int_{-T/2}^{T/2} f(t) . \exp(-jm\omega_1 t) . dt$$

In the new situation where we let the period tend to infinity, each individual coefficient becomes vanishingly small, and it might seem that the above formulae are no longer useful. However the product $a_m . T$ does not vanish as $T \to \infty$, so we now choose to write this as a new variable G. Furthermore, as $T \to \infty$, $\omega_1 \to 0$, and the term $m\omega_1$ tends to a continuous rather than a discrete variable which we will denote by ω. Since the variable G is a function of this continuous frequency variable ω, we now rewrite the second of the above equations as

$$G(\omega) = \int_{-\infty}^{\infty} f(t) . e^{-j\omega t} . dt$$

Returning to the first equation which expresses $f(t)$ as a sum of an infinite set of harmonic components, we now have

$$f(t) = \sum_{m=-\infty}^{\infty} \frac{G(\omega)}{T} \exp(jm\omega_1 t) = \sum_{m=-\infty}^{\infty} G(\omega) . \frac{\omega_1}{2\pi} . \exp(jm\omega_1 t)$$

Once again the term $m\omega_1$ is replaced by the continuous variable ω, and the fundamental frequency ω_1 (which is now vanishingly small) is written as $d\omega$. Our summation becomes in the limit an integration and the equation is thus rewritten as

$$f(t) = \frac{1}{2\pi} \int_{-\infty}^{\infty} G(\omega) . e^{j\omega t} . d\omega$$

These two derived equations, which show how a non-repetitive time-domain waveform is related to its continuous spectrum, are known as the Fourier integral equations. They are of such central importance in the analysis of signals that they are now repeated

$$G(\omega) = \int_{-\infty}^{\infty} f(t) . e^{-j\omega t} . dt$$

$$f(t) = \frac{1}{2\pi} \int_{-\infty}^{\infty} G(\omega) . e^{j\omega t} . d\omega$$

It is very important to grasp the significance of these two equations. The first tells us how the energy of the waveform $f(t)$ is continuously distributed in the frequency range between $\omega = \pm\infty$, whereas the second shows how, in effect, the waveform may be synthesised from an infinite set of exponential functions of the form $e^{j\omega t}$, each weighted by the relevant value of $G(\omega)$.

It is perhaps worth exploring a little further the physical meaning of the continuous frequency variable $G(\omega)$. It is indeed hard to visualise a wave such as an isolated pulse being composed of an infinite set of waves of infinitely small amplitudes, or the energy of such a waveform being continuously distributed in the frequency domain. The task is perhaps made easier by referring to the more familiar situations illustrated in figure 3.4. Figure 3.4(a) shows a simply-supported beam loaded at a number of distinct points, whereas in figure 3.4(b) the beam is

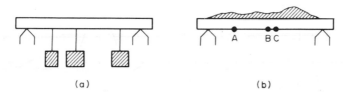

(a)   (b)

Figure 3.4   *Two types of beam-loading. In (a) the beam is loaded at three distinct points, whereas in (b) it is loaded continuously along its length*

continuously loaded along its length by, say, gravel or concrete. In the first case it is easy to say that the loading is applied only at certain discrete points, just as a repetitive signal waveform contains only certain discrete frequencies. However, if one is asked what the load on the continuously-loaded beam is at a point such as A, the answer must be that at that point (or any other) the applied load is vanishingly small. The sensible approach is to ask what the average loading is over a small distance such as B–C, and to give the answer in kilograms per metre. In the same way a continuous frequency spectrum implies that the component at any point-frequency is vanishingly small and that it is only sensible to ask about the energy contained in a small band of frequencies centred around that point. Therefore the variable $G(\omega)$ is best thought of as a frequency density function.

3.3.2 Examples of continuous spectra

To illustrate the use of the Fourier integral, we now formally evaluate the spectrum of the isolated pulse waveform of figure 3.5(a). The limits of integration are clearly reduced so that

$$G(\omega) = \int_{-\tau}^{\tau} 1 \cdot e^{-j\omega t}\, dt = -\frac{1}{j\omega}\left[e^{-j\omega t} \right]_{-\tau}^{\tau}$$

$$= -\frac{1}{j\omega}(e^{-j\omega\tau} - e^{j\omega\tau}) = \frac{2\sin\omega\tau}{\omega}$$

$$= 2\tau\left(\frac{\sin\omega\tau}{\omega\tau} \right)$$

As would be expected from our earlier discussion, this function is of $(\sin x)/x$ form and is illustrated in figure 3.5(b). It passes through zero whenever $\sin \omega\tau = 0$, which occurs when ω is an integer multiple of (π/τ) radians/second. It may seem strange that the pulse contains no energy at such frequencies, but it is not hard to demonstrate. Consider, for example, the frequency $\omega = \pi/\tau$, or $f = 1/2\tau$. If we wish to find out how much of such a frequency is contained in the pulse, the rule is to multiply the pulse by the appropriate sinusoidal waveform and to integrate over the interval of interest. It is clear that the result must be zero because the integral of the sinusoidal waveform over any interval of 2τ is always zero. Thus the pulse contains no energy at the frequency $f = 1/2\tau$ hertz.

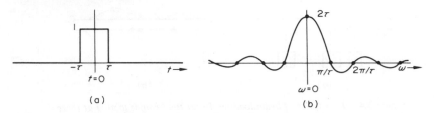

Figure 3.5   *(a) A single isolated pulse and (b) its frequency spectrum*

The spectrum of this pulse waveform illustrates a number of important points about time-limited signals. If the pulse duration (2τ) is very large, its spectral energy is concentrated around $\omega = 0$, and in the limit the $(\sin x)/x$ function becomes just a line at zero frequency; the pulse waveform has become, in other words, just a steady (d.c.) level of infinite duration. Conversely, if the pulse is made extremely short, higher frequencies are increasingly represented in its spectrum, and in the limiting case of an infinitely narrow pulse the spectrum becomes flat and extends throughout the frequency band. These results are shown in figure 3.6, and underline the important principle that a waveform which is very time-limited (that is, of short duration in the time-domain) will occupy a wide band of frequencies,

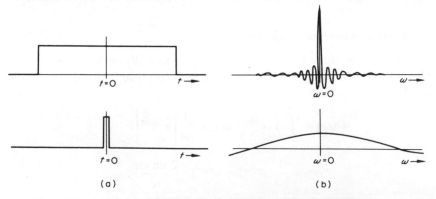

Figure 3.6   *(a) Two isolated pulses, one of long and the other of short duration, and (b) their corresponding spectra. Only a small part of the (sin x /x) spectrum of the short pulse is shown*

and vice versa. Finally these results suggest what to expect when a nominally continuous or repetitive signal is switched on and off. The above pulse waveforms may indeed be regarded as steady 'signals' which are turned on and off. The sudden changes which occur at the moments of switching introduce, in effect, new frequencies, which cause the single line spectrum of a steady level to be broadened. It is important to appreciate these properties of a time-limited signal, whether it represents an electrical waveform in a communications channel or data recorded or observed over a finite interval. The time-limitation or truncation of signals will be further discussed in chapter 8.

Having defined the Fourier transform equations it would of course be possible to evaluate the spectrum of various non-repetitive signals. The difficulty of the task depends upon the form of the integral to be evaluated and therefore varies greatly according to the signal waveform chosen. Here we must content ourselves with one further example which aids an understanding of continuous spectra, the

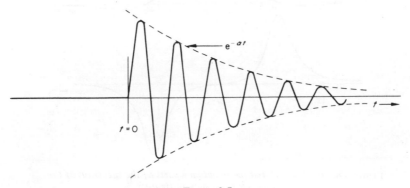

Figure 3.7

exponentially decaying sinusoidal waveform of figure 3.7. Analytically, this waveform is described by

$$f(t) = e^{-\alpha t} \sin \omega_0 t, \qquad t \geqslant 0$$
$$= 0, \qquad\qquad t < 0$$

Its spectrum is therefore given by

$$g(\omega) = \int_{-\infty}^{\infty} f(t) . e^{-j\omega t} . dt = \int_{0}^{\infty} e^{-\alpha t} . \sin \omega_0 t . e^{-j\omega t} . dt$$

The integration is quite straightforward if $\sin \omega_0 t$ is replaced by $(1/2j)[\exp(j\omega_0 t) - \exp(-j\omega_0 t)]$, and yields

$$G(\omega) = \frac{\omega_0}{\alpha^2 + \omega_0^2 - \omega^2 + j2\alpha\omega}$$

The first thing to notice about this result is that $G(\omega)$ is complex, whereas in the case of the isolated pulse of figure 3.5 it is purely real. The reason for this difference is that the pulse waveform is drawn symmetrical about $t = 0$, so that its spectrum contains only cosine components. The decaying sinusoidal wave, however, is far from symmetrical about $t = 0$ and could therefore only be synthesised from both sines and cosines, and this fact is reflected by its complex spectrum. In figure 3.8 the real and imaginary parts of its spectrum are sketched for the case when α is small compared with ω_0, so that the rate of decay of the sinusoid is relatively slow, as in figure 3.7. The imaginary part of $G(\omega)$ peaks strongly in the region of $\omega = \omega_0$, which is not surprising since the waveform $f(t) = e^{-\alpha t} \sin \omega_0 t$ must be expected to contain large sinusoidal components at frequencies close to ω_0. On the other hand we would not anticipate strong cosine components in this frequency region and the form of the real part of $G(\omega)$ confirms this view. Two further points are worth mentioning: firstly, the real

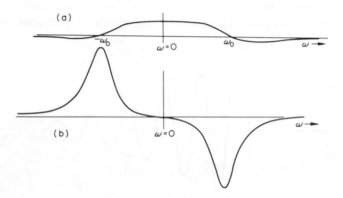

Figure 3.8   *(a) The real and (b) imaginary parts of the spectrum of the waveform shown in figure 3.7*

part of $G(\omega)$ is an even function of ω, whereas the imaginary part is odd. This is because the signal waveform is a real function of time which may be represented as the sum of cosine and sine components. Indeed, a spectrum $G(\omega)$ may invariably be expressed[16] as the sum of two other functions of ω as follows

$$G(\omega) = a(\omega) + jb(\omega)$$

where a is an even function and b is an odd function. The second point is that since

$$\sin x = \frac{1}{2j}(e^{jx} - e^{-jx}) = -\frac{j}{2}(e^{jx} - e^{-jx})$$

a positive sine component $\sin \omega_0 t$ in a signal gives rise to a negative term in the imaginary part of its spectrum at frequency ω_0, and a positive term at frequency $-\omega_0$. This accounts for the negative values of the curve of figure 3.8(b) at positive values of ω.

3.3.3 Symmetry of the Fourier integral equations

There is an obvious symmetry in the two Fourier integral equations

$$G(\omega) = \int_{-\infty}^{\infty} f(t) \cdot e^{-j\omega t} \cdot dt$$

$$f(t) = \frac{1}{2\pi} \int_{-\infty}^{\infty} G(\omega) \cdot e^{j\omega t} \cdot d\omega.$$

Indeed, apart from the $(1/2\pi)$ multiplier in the second equation (which arises from the use of angular frequency ω rather than frequency expressed in cycles per second, or hertz) and the change of sign in the exponential index, the equations are identical in form. The symmetry between time and frequency domains becomes perfect if we consider an even time function such as the isolated pulse waveform already discussed (figure 3.5), which has an even spectrum containing only cosines. In such a case, if t' is substituted for $(-t)$ the first equation becomes

$$G(\omega) = \int_{\infty}^{-\infty} f(-t') \cdot e^{j\omega t'} (-dt') = \int_{-\infty}^{\infty} f(-t') \cdot e^{j\omega t'} \cdot dt'$$

but if $f(t)$ is an even function, $f(-t') = f(t')$, and therefore

$$G(\omega) = \int_{-\infty}^{\infty} f(t') \cdot e^{j\omega t'} \cdot dt'$$

which is now identical in form to the second equation apart from the $(1/2\pi)$ multiplier. The corollary is that, since the spectrum of the square pulse waveform is of $(\sin x)/x$ form, a time function with a $(\sin x)/x$ shape has spectral energy evenly distributed in a certain range $\pm \omega'$, and none outside it. This symmetry between time and frequency domains is illustrated in figure 3.9, and accounts for the common description of the Fourier integral equations as the 'Fourier transform pair'.

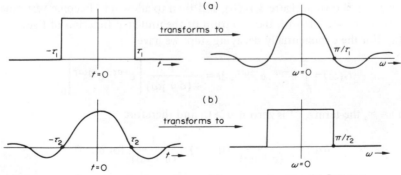

Figure 3.9 *Symmetry between time and frequency domains. (a) Just as a rectangular pulse waveform has a (sin x/x) spectrum, so (b) a time function of (sin x/x) form has a rectangular spectrum, with energy equally distributed in the range $-\pi/\tau_2 < \omega < \pi/\tau_2$*

3.3.4 Limitations of the Fourier transform

So far it has been implied that the Fourier transform may be successfully used for any non-repetitive signal, but there are in fact limitations and difficulties in its application. An important limitation may be inferred from the equation

$$G(\omega) = \int_{-\infty}^{\infty} f(t) \cdot e^{-j\omega t} \, dt$$

The Fourier transform will clearly only exist if the right-hand side of the equation is finite. Furthermore, since the magnitude of $e^{-j\omega t}$ is always unity it is sufficient for the transform to exist if

$$\int_{-\infty}^{\infty} |f(t)| \cdot dt < \infty$$

Various waveforms of practical interest, such as continuous sine and cosine functions and the so-called unit step function illustrated in figure 3.10(a) do not

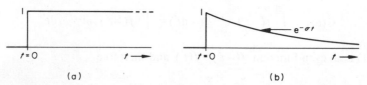

(a)   (b)

Figure 3.10   *(a) The unit step function, and (b) an exponentially decaying step function*

meet this latter condition, and strictly speaking do not therefore possess Fourier transforms. Fortunately, however, it is possible to consider them as limiting cases of waveforms which do possess transforms. For example, we may evaluate the Fourier transform of a sine or cosine wave which exists only in the interval $-T < t < T$: we then allow T to become very large and in the limit approach the transform of the continuous function itself[2]. In the case of the unit step function, it is convenient first to evaluate the Fourier transform of the exponentially decaying step shown in figure 3.10(b), and then to allow σ to become very small: in the limit, we are left with the spectrum of the unit step function of figure 3.10(a). For the exponentially decaying step, we have

$$G(\omega) = \int_{0}^{\infty} e^{-\sigma t} \cdot e^{-j\omega t} \cdot dt = \frac{1}{-(\sigma + j\omega)} \left[e^{-\sigma t} \cdot e^{-j\omega t} \right]_{0}^{\infty}$$

when $t = \infty$, the term $e^{-\sigma t}$ is zero if $\sigma > 0$, and therefore

$$G(\omega) = \frac{1}{-(\sigma + j\omega)} \cdot (0 - 1) = \frac{1}{(\sigma + j\omega)}, \text{ for } \sigma > 0$$

Before deriving the transform of the step function itself by letting σ tend to zero, it is appropriate to mention the problem of singularities[2,9]. It quite often occurs that the value of a function is undefined at a certain point (or points): such a point is referred to as a singularity. If we let $\sigma \to 0$ in the above example, the value of $G(\omega) \to \infty$ at $\omega = 0$ and there is therefore a singularity at $\omega = 0$. The step function waveform itself (see figure 3.10(a)) is not precisely defined at $t = 0$, since there is a discontinuity at this point. In the next chapter we shall make extensive use of another singularity function, the unit impulse, for the mathematical description of sampled-data signals.

Singularities therefore occur quite often in the context of signal theory. From a mathematical point of view, the presence of such singularities may or may not give rise to special difficulties, but in any case caution should be exercised, particularly when considering the behaviour of the function in the region of a singularity.

Returning to the question of the Fourier transform of the unit step function, we now let $\sigma \to 0$ in the above expression. At first sight $G(\omega)$ would appear to tend to $(1/j\omega)$, but special care must be taken because of the singularity at $\omega = 0$. Suppose we make σ small but not zero. Over the range of frequencies for which $\omega \gg \sigma$, $G(\omega) \approx (1/j\omega)$; but when ω is virtually zero so that $\omega \ll \sigma$, we have $G(\omega) \approx (1/\sigma)$. In the limit as σ becomes smaller and smaller, we are left with a continuous spectrum equal to $(1/j\omega)$, together with a very large spike of amplitude $1/\sigma$ centred on $\omega = 0$. The latter is really a spectral line representing the d.c. or average value of the step waveform. Thus the unit step function has a spectrum consisting of an infinite set of sinusoidal components of amplitude inversely proportional to frequency, together with a zero-frequency component representing the average value of the waveform (which, over the interval $-\infty < t < \infty$, is 0·5).

In effect, we have derived the spectrum of the unit step function by multiplying it by a factor $e^{-\sigma t}$, which makes the Fourier integral convergent. This useful technique may be applied to other waveforms, although such a 'convergence factor' will not work for a signal waveform which extends throughout all time, because the value of $e^{-\sigma t}$, with σ positive, increases without limit as t becomes more and more negative. Therefore the method may only be successfully applied to a waveform which is 'switched on' at some definite instant and may be considered zero beforehand; normally the instant at which the signal first assumes a nonzero value is designated $t = 0$. If this convergence factor is applied to the general case of a signal $f(t)$ which is zero before $t = 0$, we may define a modified version of the Fourier transform

$$G_1(\omega) = \int_0^\infty f(t) . e^{-\sigma t} . e^{-j\omega t} . dt = \int_0^\infty f(t) . e^{-(\sigma + j\omega)t} . dt$$

Strictly speaking, of course, G_1 is a function not only of ω, but also of the factor σ, which must be chosen so that the integral converges in any particular case. For convenience we introduce the new variable $s = (\sigma + j\omega)$, and we therefore write G_1 as a function of s

$$G_1(s) = \int_0^\infty f(t) e^{-st} . dt$$

In doing so, we have defined the Laplace transform of the signal $f(t)$.

3.4 The Laplace transform

Born some twenty years before Fourier in Normandy, Pierre Simon Marquis de Laplace became perhaps the greatest theoretical astronomer since Newton. Although he is best remembered for contributions to the understanding of planetary motion, the mathematical basis of his work has found widespread application in other fields.

3.4.1 Relationship with the Fourier transform

The Laplace and Fourier transforms are closely related. As we have seen, the Fourier transform allows a signal to be expressed as a sum of sinusoidal and cosinusoidal components which exist over all time, past, present and future, each component being represented by a pair of imaginary exponential terms of the form $e^{j\omega t}$. By introducing a so-called convergence factor it is possible to derive the frequency spectra of certain signals for which the Fourier integral may not otherwise be evaluated. So far we have thought of this convergence factor $e^{-\sigma t}$ as being applied as a multiplier to an awkward signal, so that the integral may be evaluated, and we then let σ tend to zero. But just as the form of the Fourier integral

$$G(\omega) = \int_{-\infty}^{\infty} f(t) . e^{-j\omega t} . dt$$

implies that $f(t)$ is being analysed as an infinite set of exponential terms of the form $e^{j\omega t}$, so the form of the Laplace transform equation

$$G(s) = \int_{0}^{\infty} f(t) . e^{-st} . dt$$

suggests that we should think of the Laplace transform as representing $f(t)$ by an infinite set of terms of the form e^{st}, where s is in general a complex number known as the 'complex frequency'. Such terms give rise to not only the sine and cosine waves of the Fourier method, but also growing and decaying sine and cosine waves and growing and decaying exponentials, typical examples of which are illustrated in figure 3.11.

Since the Laplace transform in effect analyses a signal into both oscillatory and non-oscillatory functions which expand and contract with time, it allows the signal $f(t)$ to be less restricted than in the case of the Fourier integral. On the other hand, there are still restrictions because the real part σ of the variable s must always be sufficient to provide convergence. If for example $f(t)$ contains a growing exponential component e^{7t}, the integral will only converge if the term $e^{-\sigma t}$ at least counteracts its growth, which means that σ must be at least 7. But a term in $f(t)$ equal to, for example, $\exp(t^2)$ must dominate $e^{-\sigma t}$ at sufficiently large values of t regardless of the value of σ, and hence the Laplace transform could not be used. The other restriction, already mentioned, is that the Laplace transform cannot

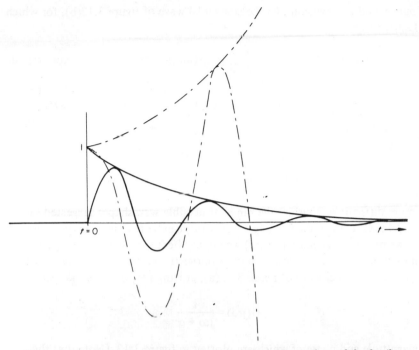

Figure 3.11   *Typical examples of waveforms which are members of the family*
defined by $f(t) = e^{st}$ for $t > 0$, when s is complex

cope with signals which extend into the infinite past, and the integral is only
evaluated in the interval $0 < t < \infty$. The reason for this is that values of σ which
ensure convergence in the interval $0 < t < \infty$ do not also give it in the interval
$-\infty < t < 0$.

3.4.2 Use of the Laplace transform

As a first example, consider the decaying exponential waveform shown in figure
3.12(a). Its Laplace transform is given by

$$G(s) = \int_0^\infty f(t) \cdot e^{-st} \, dt = \int_0^\infty e^{-(s+\alpha)t} \, dt$$

$$= -\frac{1}{(s+\alpha)} \left[e^{-(s+\alpha)t} \right]_0^\infty = \frac{1}{s+\alpha}$$

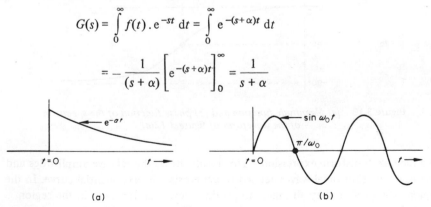

(a)   (b)

Figure 3.12

For our second case we consider the sinusoidal wave of figure 3.12(b), for which the Laplace transform is

$$G(s) = \int_0^\infty \sin \omega_0 t \cdot \exp(-st)\, dt = \int_0^\infty \frac{1}{2j} \left[\exp(j\omega_0 t) - \exp(-j\omega_0 t)\right] \exp(-st) \cdot dt$$

$$= \frac{1}{2j} \left[-\frac{1}{(s - j\omega_0)} \exp\left[-(s - j\omega_0)t\right] + \frac{1}{(s + j\omega_0)} \exp\left[-(s + j\omega_0)t\right] \right]_0^\infty$$

$$= \frac{1}{2j} \left(\frac{1}{(s - j\omega_0)} - \frac{1}{(s + j\omega_0)} \right) = \frac{\omega_0}{(s + j\omega_0)(s - j\omega_0)}$$

It is not easy to appreciate these results, because although one can fairly easily imagine a signal being synthesised from a number of continuous sinusoidal waves, it is more difficult to consider the range of possible waveshapes suggested by figure 3.11. Of course, the variable $s = (\sigma + j\omega)$ can be real, imaginary, or complex, and if we make it imaginary ($s = j\omega$) we are in effect considering the signal to be composed of sinusoidal waves, as in the Fourier transform. In the case of the decaying exponential wave of figure 3.12(a), putting $s = j\omega$ yields the spectrum

$$G(j\omega) = \frac{1}{j\omega + \alpha}$$

the magnitude and phase of which are plotted in figure 3.13. (Note that the variable $G(\omega)$ is often written as $G(j\omega)$, denoting that s has been put equal to $j\omega$

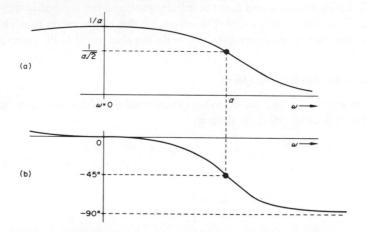

Figure 3.13   *(a) Magnitude function and (b) phase function of the spectrum of the waveform of figure 3.12(a)*

in a Laplace transform expression.) This result shows the relative amplitudes and phases of the sinusoidal waves needed to synthesise the exponential curve. In the region $-\infty < t < 0$, they all cancel to produce zero resultant, but in the region $0 < t < \infty$ they add together to give the required waveform. If we put $s = j\omega$ in order to investigate the spectrum of the waveform of figure 3.12(b) we get the following expression

$$G(j\omega) = \frac{\omega_0}{(j\omega + j\omega_0)(j\omega - j\omega_0)}$$

In this case $G(j\omega)$ displays singularities which imply that the magnitude of the frequency spectrum is infinite at $\omega = \pm\omega_0$, whereas we have in fact chosen a sinusoidal wave of unit amplitude as our time function. This difficulty arises because by putting $s = j\omega$ we are effectively trying to evaluate the Fourier transform of a wave which continues for ever, for which the integral is not convergent. In such cases the results of substituting $j\omega$ for s in a Laplace transform expression are not easily interpreted.

So far we have shown how to derive the Laplace transform of a time function, expressing it as a function of the complex frequency variable s, but no mention has been made of the opposite process of deriving the time function corresponding to a given function of s. One of the main difficulties of the Laplace transform method is that this reverse process of 'inverse transformation' is not a straight-forward mathematical operation. Formally, the inverse transform is defined as

$$f(t) = \frac{1}{2\pi j} \int_{\sigma - j\infty}^{\sigma + j\infty} G(s) \cdot e^{st} \cdot ds$$

and while the general form of this equation is similar to the corresponding Fourier integral equation, the limits of $(\sigma + j\infty)$ give rise to considerable difficulty in practice. The evaluation of this integral requires familiarity with functions of complex variables and the calculus of residues[2,8]. The approach adopted here is the practical one commonly used in texts which deal with the Laplace transform, namely to provide a table of some of the more common transform pairs for ready reference. Such a table is given at the end of the text.

Even an extensive table of Laplace transforms can hardly hope to cover all the functions likely to be encountered in practice, and while it is beyond the scope of this book to investigate the various techniques of inverse transformation, one simple procedure which is quite often useful will now be described. It relies upon the fact that the inverse Laplace transform of the sum of a number of functions is equal to the sum of their inverse transforms, and is called the *method of partial fractions*. In a great many cases, it is found that a function of s (of which the inverse transformation is required) consists of products of simple linear factors, and the method of partial fractions allows us to express such a function as the sum of a number of simpler functions, each of which appears in our table of transforms. To take a straightforward example, suppose we have

$$G(s) = \frac{(s + \alpha)}{(s + \beta)(s + \gamma)}$$

We write the identity

$$\frac{A}{(s+\beta)} + \frac{B}{(s+\gamma)} = \frac{(s+\alpha)}{(s+\beta)(s+\gamma)}$$

and multiply both sides by $(s+\beta)(s+\gamma)$, giving

$$A(s+\gamma) + B(s+\beta) = (s+\alpha)$$

Equating coefficients of equal powers of s yields

$$A + B = 1 \qquad \text{and} \qquad A\gamma + B\beta = \alpha$$

from which

$$B = \frac{(\gamma - \alpha)}{(\gamma - \beta)} \qquad \text{and} \qquad A = \frac{(\alpha - \beta)}{(\gamma - \beta)}$$

We now have $G(s)$ expressed in a much simpler form; both $A/(s+\beta)$ and $B/(s+\gamma)$ have an inverse transform of simple exponential form, so the time function corresponding to $G(s)$ is equal to the sum of two exponentials. The partial fraction method applies to any $G(s)$ which may be expressed as the ratio of two polynomials in s, provided that numerator polynomial is of smaller degree than the denominator polynomial; for a more comprehensive treatment the reader is referred to almost any book on advanced algebra.

It might be objected that the Laplace transform is hardly worth the trouble, if its ability to cope with a few signal waveforms not amenable to the Fourier transform is offset by the difficulty of inverse transformation and by the greater conceptual problems presented by complex frequencies. It should however be stressed that this elementary introduction to the Laplace transform does little to suggest the full value of the method. This will become clearer in chapter 7, where further aspects of the Laplace transform are discussed in the context of signal processing. Suffice it to say at this stage that the Laplace transform is a powerful mathematical tool which may be used to solve a great many problems other than those of signal analysis. In particular (and this matter will also be discussed again in chapter 7) a problem stated as a set of differential equations may often be reduced to a set of much simpler algebraic equations if the Laplace transform is used. Since the dynamics of certain types of mechanical and electrical systems (of which many are of great interest for signal processing) may be expressed mathematically in differential equation form, the Laplace transform method can be of great value in exploring their performance. With this possibility in mind, we now investigate some further aspects of the Laplace transform which will prove useful for subsequent work.

3.4.3 The pole—zero description of a signal

Consider once again the function

$$G(s) = \frac{1}{s+\alpha}$$

which, as we have already seen in the previous section, represents the Laplace transform of a decaying exponential signal waveform. It is a general property of such functions that there are values of the complex frequency variable s which make $G(s)$ tend to infinity and values of s which make $G(s)$ zero; such values are known as the 'poles' and 'zeros' of $G(s)$ respectively, and they give rise to a convenient graphical description of the corresponding signal. In the simple case quoted above, it is clear that $G(s) \rightarrow \infty$ if $s = -\alpha$, so the latter value is a pole of $G(s)$; there are no zeros (apart from the trivial case of $s = \infty$). If we consider a more complicated function of the general form

$$G(s) = \frac{A(s - z_1)(s - z_2)(s - z_3) \dots}{(s - p_1)(s - p_2)(s - p_3) \dots}$$

where A is a constant, then the values $s = z_1, z_2, z_3$, etc., represent zeros of $G(s)$ and $s = p_1, p_2, p_3$, etc., represent poles. It is indeed always possible to write a Laplace transform in the above way, and therefore to define the signal in terms of a set of poles and zeros. It is found[17] that poles are either real, or they occur in complex conjugate pairs, and the same is true of zeros. Thus a pole at $s = \alpha + j\beta$ is matched by one at $s = \alpha - j\beta$, and this must always be the case when the corresponding signal is a real function of time.

It is common practice to represent the poles and zeros of $G(s)$ graphically by drawing their positions on a so-called Argand diagram (complex plane). In such a diagram the real part of a complex variable is plotted along the abscissa, and the imaginary part along the ordinate. The poles and zeros of a function $G(s)$ are in general complex values of s, so the Argand diagram gives a convenient method of displaying them, in which case it is widely referred to as the 's-plane' diagram. Suppose, for example, we have a time function $f(t)$, the Laplace transform of which is

$$G(s) = \frac{4(s^2 - 2s)}{s^2 + 2s + 10} = \frac{4(s)(s - 2)}{(s + 1 + j3)(s + 1 - j3)}$$

Apart from the constant multiplier of 4, we may completely represent the function $G(s)$ by drawing zeros at $s = 0$ and $s = 2$, and poles at $s = -1 \pm j3$, in the complex plane. These are shown plotted in figure 3.14, using the normally adopted symbols for the poles and zeros.

Quite apart from the convenience of describing a signal in terms of a set of s-plane poles and zeros, the method also has the advantage of allowing the form of the Fourier frequency spectrum (in which the signal is considered as the sum of a set of continuous sinusoidal waves) to be readily visualised[17]. We have already seen how, if s is replaced by $j\omega$ in a Laplace transform expression, we derive the Fourier spectrum of the corresponding signal. To take again the simple case of the decaying-exponential signal of figure 3.12(a), its Laplace transform

$$G(s) = \frac{1}{(s + \alpha)}$$

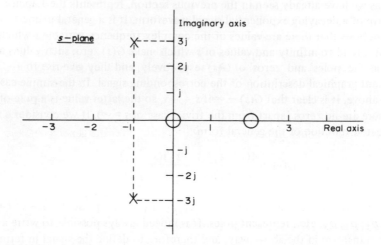

Figure 3.14   *The poles and zeros of a typical Laplace transform function G(s)*

becomes its Fourier frequency spectrum

$$G(j\omega) = \frac{1}{(j\omega + \alpha)}$$

when s is replaced by $j\omega$. Now the pole-zero configuration of this signal consists of a single pole at $s = -\alpha$, and hence the denominator term $(j\omega + \alpha)$ may be represented by a vector in the s-plane drawn from the pole to a point on the imaginary axis representing a particular value of ω, as shown in figure 3.15. The length of this vector represents the magnitude of the term $(j\omega + \alpha)$ and the angle it makes with the positive real axis represents its phase. Of course, the term

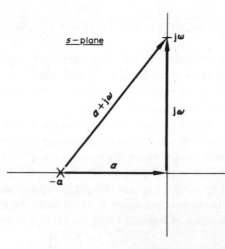

Figure 3.15   *A single pole at $s = -\alpha$, and the three s-plane vectors α, $j\omega$, and*
$(\alpha + j\omega)$

($j\omega + \alpha$) occurs in the denominator of $G(j\omega)$, so the magnitude of $G(j\omega)$ is equal
to the reciprocal of the length of the vector ($j\omega + \alpha$), associated with a phase
angle equal but opposite in sign to the vector's phase. In this particular case it is
clear that when $\omega = 0$ the vector ($j\omega + \alpha$) has a length of α, and a phase angle of
zero, whereas as ω becomes very large so does its length and the phase angle tends
to 90° ($\pi/2$ radians). The corresponding frequency spectrum therefore has the
magnitude $1/\alpha$ at $\omega = 0$, tending to zero as $\omega \to \infty$, and has a phase term which
varies between 0° and $-90°$ (see figure 3.13).

With a little practice, it becomes quite simple to infer the general shape of the
Fourier spectrum (normally referred to simply as 'the spectrum') of a signal
waveform, given the s-plane pole-zero configuration. When, as is the general case,
there are a number of poles and zeros, the magnitude of the spectrum at a
particular frequency is found by taking the product of the lengths of the various
vectors drawn from the zeros to the relevant point on the imaginary axis (the
'zero vectors'), and dividing this by the product of the lengths of the vectors
drawn from the poles (the 'pole vectors'). Similarly, the net phase term is found as
the sum of the phases of the various zero vectors minus the sum of the phases of
the pole vectors. Bearing this rule in mind, it is not hard to see that a pole close
to the imaginary axis in the s-plane gives rise to a hump in the frequency spectrum
as ω approaches a value equal to the pole's imaginary part (that is, a pole at
$s = \alpha + j\beta$ produces a hump in the spectrum in the region of $\omega = \beta$, if α is sufficiently
small), whereas a zero in a similar position will give rise to a trough, or minimum.
In other words, the positions of poles and zeros in the s-plane indicate which
sinusoidal frequency ranges are most significantly represented in the spectrum of a
signal. These matters are illustrated further by figure 3.16, which shows two rather

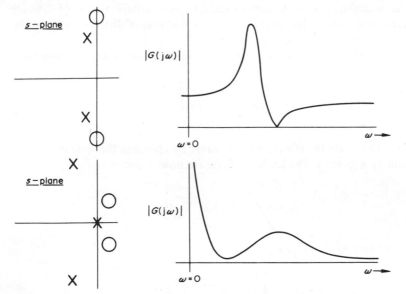

Figure 3.16   *Two pole–zero configurations and their corresponding spectral
magnitude characteristics*

more complicated pole-zero configurations and sketches of the corresponding spectral magnitude functions. It is worth noting that a pole in the right-hand half of the s-plane represents a time function which continues to grow without limit. Since such functions are not encountered in the real physical world, it must always be expected that poles will be restricted to the left-hand half of the s-plane; the same restriction does not, however, apply to zeros. Further pole-zero configurations are illustrated in the table of Laplace transforms which follows the text.

3.4.4 Some further properties

So far, our discussion has concentrated on deriving the Laplace transform of a signal waveform and its representation by a set of s-plane poles and zeros. We have two equally valid descriptions of a signal: its time-domain waveform and its frequency-domain spectrum. It is very interesting, and useful for our subsequent work, to investigate the effects in one domain caused by various types of operation performed on the function in the other domain: for example, what happens to its Laplace transform when a function of time is time-shifted, or differentiated, or integrated? Although the discussion which follows is specifically related to the Laplace transform, a comparable set of properties applies to the Fourier transform, if the variable s is replaced by $j\omega$.[1,2] This is not surprising if we regard the Laplace transform as a generalisation of the Fourier transform. Furthermore, since there is a certain symmetry in the Fourier transform equations, we should expect this symmetry to be reflected in the properties. Broadly speaking, whether we are dealing with the Laplace or Fourier transform, we might expect that the effect on the frequency-domain function caused by, say, differentiation of the time function is similar to the effect on the time function of differentiation in the frequency domain. This is indeed the case[2,9].

Change of scale Assuming that the Laplace transform $G(s)$ of a time function $f(t)$ is given by

$$G(s) = \int_0^\infty f(t) . e^{-st} . dt$$

we first investigate the effect on $G(s)$ caused by changing the scale in the time-domain by a factor a. The Laplace transform now becomes

$$G_1(s) = \int_0^\infty f(at) . e^{-st} . dt$$

Replacing (at) by t', we have

$$G_1(s) = \frac{1}{a} \int_0^\infty f(t') . e^{-(s/a). t'} . dt = \frac{1}{a} . G(s/a)$$

Time-shift Next, suppose that the function $f(t)$ is shifted forward in time by an amount T_0. Whereas it was initially zero for $t < 0$, it is now zero for $t < T_0$. Hence its Laplace transform is given by

$$G_2(s) = \int_{T_0}^{\infty} f(t - T_0)\, e^{-st}.\, dt$$

Replacing $(t - T_0)$ by t', we have

$$G_2(s) = \int_0^{\infty} f(t').\exp\{-s(t' + T_0)\}.\, dt' = \exp(-sT_0).\int_0^{\infty} f(t').\exp(-st').\, dt'$$

$$= \exp(-sT_0).\, G(s)$$

Frequency-shift Suppose next that the time function $f(t)$ is multiplied by the exponential term e^{-at}. Its Laplace transform now becomes

$$G_3(s) = \int_0^{\infty} \{e^{-at}.\, f(t)\}.\, e^{-st}.\, dt$$

$$= \int_0^{\infty} f(t).\, e^{-(s+a)t}.\, dt = G(s + a)$$

Time differentiation The Laplace transform of the first derivative of $f(t)$ is given by

$$G_4(s) = \int_0^{\infty} \frac{d}{dt}\{f(t)\}.\, e^{-st}.\, dt$$

which may be integrated by parts to give

$$G_4(s) = \left[f(t).\, e^{-st} \right]_0^{\infty} + s \int_0^{\infty} f(t).\, e^{-st}.\, dt$$

$$= s.\, G(s) - f(0)$$

where $f(0)$ represents the value of $f(t)$ immediately prior to $t = 0$. Since we are generally concerned with functions which are zero before $t = 0$, the term $f(0)$ may normally be disregarded, giving

$$G_4(s) = s.\, G(s)$$

However, for generality, we should retain the term $f(0)$. The reasons for this will become clearer in chapter 7 when we deal with the response of linear systems which have energy stored in them prior to $t = 0$.

In a similar manner, the transform of the nth derivative of $f(t)$ is given by

$$G_5(s) = s^n.\, G(s) - s^{n-1}f(0) - s^{n-2}\frac{df(0)}{dt} - \cdots \frac{df^{n-1}(0)}{dt^{n-1}}$$

where all terms $f(0)$, $df(0)/dt$, . . . represent values of $f(t)$ and its derivatives immediately prior to $t = 0$.

Time integration If $f(t)$ is integrated between the limits $t = 0$ and $t = t$, its Laplace transform becomes

$$G_6(s) = \int_0^\infty \left(\int_0^t f(t) \cdot dt \right) \cdot e^{-st} \cdot dt$$

This may be integrated by parts to give

$$G_6(s) = \left[\frac{-e^{-st}}{s} \int_0^t f(t) \cdot dt \right]_0^\infty + \frac{1}{s} \int_0^\infty f(t) \cdot e^{-st} \cdot dt$$

The first term on the right-hand side of this equation is zero, so that

$$G_6(s) = \frac{1}{s} \cdot G(s)$$

This table of properties could be extended to include such operations as frequency-domain differentiation and integration. However we have now covered a number of cases which will prove particularly useful in the remainder of this text. From our point of view, the main lesson is that such operations as time-shifting, differentiation, and integration cause definite and quite straightforward modifications to the frequency-domain description of a signal. For example, the Laplace transform of the first derivative of $f(t)$ is simply equal to s times that of $f(t)$ itself (less a constant equal to $f(0)$): the transform of the integral of $f(t)$ is found by dividing that of $f(t)$ by s, and so on. Various references[2,9] give tables and other properties of the Laplace transform.

Problems

1. Using considerations of symmetry, what can be predicted about the coefficients (a_m) of the exponential Fourier series for the repetitive waveforms illustrated in figures 2.4 and 2.10?

2. Expand the repetitive function illustrated as an exponential Fourier series, finding terms up to and including the third harmonic.

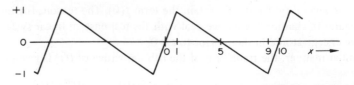

3. Derive the exponential Fourier series for a repetitive pulse waveform in which

the period is ten times as great as the pulse duration, and the pulse height is unity. Assume that

(i) the instant $t = 0$ coincides with the centre of one of the pulses,
(ii) the instant $t = 0$ coincides with the leading edge of one of the pulses.

4. Derive and sketch the Fourier spectra of the two non-repetitive waveshapes illustrated. Are the spectra real, imaginary, or complex, and why?

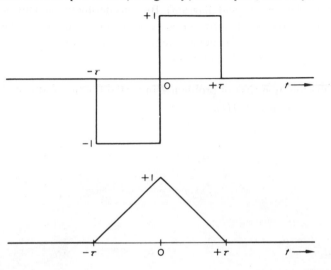

5. Find the Fourier spectra of the single-sided and double-sided decaying exponential waveforms shown. Denoting these by $G_1(\omega)$ and $G_2(\omega)$ respectively, show that the spectra are related by the formula $G_2(\omega) = 2a_1(\omega)$ where $a_1(\omega)$ is the real part of $G_1(\omega)$.

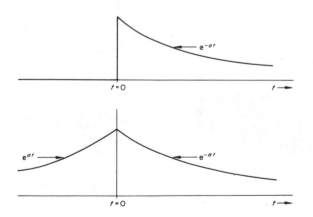

6. Derive the Laplace transform of the signal $f(t) = e^{-\alpha t} \cos \omega_0 t$ for $t > 0$, and sketch its real and imaginary parts for the case where $\omega_0 \gg \alpha$. Compare the

result with that for the decaying sinusoidal waveform shown in figure 3.7, and explain the differences.

7. Plot the poles and zeros of the function:

$$G(s) = \frac{s + 1}{s^2 + 2s + 26}$$

on a s-plane diagram (Argand diagram). By considering the lengths of vectors drawn from the poles and zeros to points on the imaginary axis, sketch the approximate form of the frequency spectrum of the corresponding time function in both magnitude and phase.

8. Express the function $G(s)$ of problem 7 in partial fraction form and hence find its inverse transform $f(t)$.

4

Sampled-Data Signals

4.1 Introduction

Sampled-data signals have defined values only at certain instants of time, and arise whenever continuous functions are measured or recorded intermittently. In recent years such signals have come to assume great importance because of developments in digital electronics and computing. Since it is not possible to feed continuous data into a digital (as opposed to analogue) computer, any signal or data input must be represented as a set of numerical values. In almost every case the numbers represent sampled values of the continuous signal at successive equally-spaced instants. An example of a sampled-data signal of this type has already been illustrated in figure 1.2, which shows successive values of midday temperature measured in a particular place. A further example, in which successive samples represent values of an electrical potential between two points in a circuit, is shown in figure 4.1.

Figure 4.1　*A sampled-data signal*

It should be emphasised from the start that a set of sample values only forms an adequate substitute for the underlying continuous signal waveform if the interval between successive samples is sufficiently small. This matter is discussed in some detail in chapter 8 where sampling and resynthesis of a continuous function from its sample values are considered as signal processing operations. The object of this chapter is limited to showing how sampled-data signals may be expressed mathematically, and how the foregoing framework of frequency-analysis techniques applies to them.

4.2   Mathematical description using the Dirac (δ) function

The Dirac function, often referred to as the impulse function, is a pulse of extremely short duration and unit area. In other words the product of its duration and its mean height is unity, even though its precise shape is undefined. The physical significance of such a function may be illustrated by an example. Suppose a mechanical impulse is delivered to a golf ball by the head of a golf club. Other things being equal, the momentum imparted to the ball and the distance it travels depend upon the value of the impulse, which is given by the product of the force and the time for which it is exerted. Or, assuming that the force is not constant, it is the area under the force–time graph which determines the impulse. Since the Dirac function is defined by its unit area, it may be used to denote a unit mechanical impulse, and we shall use the same function later in this text to describe a sudden electrical disturbance applied to a signal-processing device.

The Dirac function is also useful for describing a sampled data signal, which may be considered to consist of a number of equally-spaced pulses of extremely short duration. It is convenient in this case to think of all the signal samples as pulses of identical duration,.with their heights proportional to the values of the signal at the relevant instants. In practice, this approach proves mathematically sound, provided the pulse duration is negligible compared with the interval between successive samples.

In order to discuss the spectrum of a sampled-data signal, it is necessary to evaluate that of the unit Dirac function. Using the convention that the symbol $\delta(t)$ represents a Dirac pulse occurring at $t = 0$, we have

$$G(j\omega) = \int_{-\infty}^{\infty} \delta(t) . e^{-j\omega t} . dt$$

To evaluate this integral we make use of the so-called 'sifting property' of the Dirac function which is illustrated in figure 4.2. Here a function $f(t)$ is shown multiplied by a unit Dirac function occurring at the instant $t = a$ which is denoted by the symbol $\delta(t - a)$. Since the area of the Dirac pulse is unity, the area under the curve representing the product $[f(t) . \delta(t - a)]$ is the value of $f(t)$ at $t = a$. When the product is integrated between $t = -\infty$ and ∞, the result is just this area. Hence the sifting property may be formally stated as follows

$$\int_{-\infty}^{\infty} f(t) . \delta(t - a) . dt = f(a)$$

The spectrum $G(j\omega)$ of a Dirac pulse at $t = 0$ is therefore simply equal to the value of $e^{-j\omega t}$ at $t = 0$, which is unity. Hence

$$G(j\omega) = 1$$

This result tells us that all frequencies are equally represented by cosine components. When a large number of cosines of similar amplitude but different frequency are added together, they tend to cancel each other out everywhere except at $t = 0$,

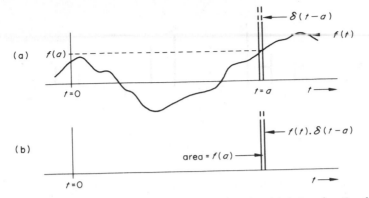

Figure 4.2   *The sifting property of the Dirac function; (a) A time function f(t)*
and a Dirac pulse of unit area occurring at t = a, and (b) the product
$$[f(t) . \delta(t - a)]$$

where they all reinforce. Therefore as higher and higher frequencies are included the resultant becomes an extremely narrow pulse centred on $t = 0$.

It is also interesting to consider the spectrum of a delayed Dirac pulse which occurs at some instant $t = T$. Clearly, such a pulse must contain the same frequency components as the one occurring at $t = 0$, except that each and every one of them is delayed by T seconds. Formally, the spectrum is given by

$$G(j\omega) = \int_{-\infty}^{\infty} \delta(t - T) . e^{-j\omega t} . dt = e^{-j\omega T}$$

The term $e^{-j\omega T}$ has unit magnitude for any value of ω, and a phase shift of $-\omega T$ radians. As we would expect, a phase shift of $-\omega T$ radians imposed upon a frequency component of ω radians/second represents a time delay of T seconds. Thus the spectrum $e^{-j\omega T}$ implies that all components are present with equal amplitude and a phase shift proportional to frequency. The same result could be obtained by noting that the Laplace transform of a unit impulse centred on $t = 0$ is also unity. Since shifting a time function by T seconds is equivalent to multiplying its Laplace transform by e^{-sT} (see section 3.4.4) the transform of a unit impulse occurring at $t = T$ is $1 . e^{-sT} = e^{-sT}$. Substitution of $j\omega$ for s gives its frequency spectrum.

4.3   Spectra of sampled-data signals

4.3.1 The discrete Fourier transform

It is now possible to write down time and frequency-domain expressions for a sampled-data signal such as that shown in figure 4.3. Denoting a sample value x

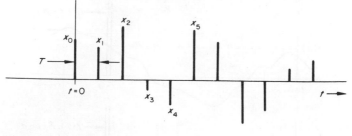

Figure 4.3

by a unit Dirac pulse weighted (that is, multiplied) by x, the signal is described by

$$f(t) = x_0 . \delta(t) + x_1 . \delta(t - T) + x_2 . \delta(t - 2T) + x_3 . \delta(t - 3T) + \ldots$$

where T is the sampling interval. Its Laplace transform may be written by inspection as

$$G(s) = x_0 + x_1 . e^{-sT} + x_2 . e^{-s2T} + x_3 . e^{-s3T} + \ldots$$

and its Fourier transform as

$$G(j\omega) = x_0 + x_1 . e^{-j\omega T} + x_2 . e^{-j\omega 2T} + x_3 . e^{-j\omega 3T} + \ldots$$

The Fourier transform of a sampled-data signal is generally referred to as a discrete Fourier transform (DFT)[12]. There are two reasons for this: firstly, the signal itself is discrete, in the sense of being defined only at discrete instants in time—the sampling instants: and secondly, the common practice of using a digital computer to evaluate the spectrum of a sampled-data signal means that its Fourier transform can only be estimated for a set of discrete values of ω. This is not to imply that $G(j\omega)$ as defined in the above equation is discrete, for it is a continuous function of ω; but with a digital computer we can only estimate it at suitably spaced intervals in ω, and use the values obtained to represent the underlying continuous function. Fortunately, as we shall show later, such a discrete representation of the spectrum need not involve any loss of essential detail.

Although it is always a simple matter to write down an expression for the spectrum of a sampled-data signal, it is less easy to visualise its form. However, some of the main features of such spectra may be inferred in the following way. Since a sampled-data signal is in effect composed of a set of weighted Dirac pulses, each of which contains energy distributed over a very wide band of frequencies, it must be expected that the signal spectrum will be similarly 'wideband'. The second point to notice is that since terms such as $e^{-j\omega T}$, $e^{-j\omega 2T}$, $e^{-j\omega 3T}$ are all repetitive in the frequency domain with a period of $2\pi/T$ radians/second, so also must be the spectrum itself. Finally, it would be reasonable to suppose that the spectrum of a sampled version of a waveform would reflect not only the frequency characteristics of the Dirac functions representing the sample pulses, but also components present in the underlying continuous signal (were this not so, the sampled version could hardly be used as a valid representation of the signal).

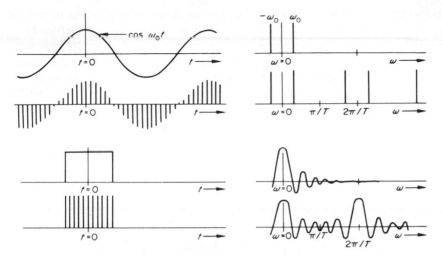

Figure 4.4 *Four signals and their spectra. The continuous and sampled version of the cosine and pulse waveforms have similar spectra in the region* $-\pi/T < \omega < \pi/T$*, but the sampled-data spectra repeat indefinitely at intervals in* ω *of* $2\pi/T$

Figure 4.4, which shows the spectra of two typical continuous signals and those of their sampled versions, confirms the above arguments. For example, the continuous wave $\cos \omega_0 t$ has two spectral lines at $\omega = \pm \omega_0$; its sampled version has a spectrum in which these two spectral lines repeat indefinitely at intervals in ω of $2\pi/T$, where T is the sampling interval. An isolated rectangular pulse has a continuous spectrum of $(\sin x/x)$ form; when sampled, the spectrum repeats itself every $2\pi/T$ radians/second. In both sampled-signal spectra, the low-frequency region between $\omega = -\pi/T$ and $\omega = \pi/T$ represents components present in the underlying continuous signal waveform, the repetitive nature of the spectra being due to the sampling process itself. In fact there is never any need to evaluate the spectrum of a sampled-data signal at frequencies above $\omega = \pi/T$, since it is bound to be merely a repetition of the portion lying between $\omega = -\pi/T$ and $\omega = \pi/T$.

In any practical situation, we are forced to consider a signal of limited duration, so that any sampled-data spectrum will be estimated from a finite set of sample values. It is interesting to consider the effect this has on the number of points on the frequency scale at which it is sensible to evaluate the DFT. To take a simple example consider the sampled-data record shown in figure 4.5(a), which consists of nine sample values. Suppose this signal is repeated end-on-end, *ad infinitum*, as indicated in figure 4.5(b). It is now a truly repetitive signal with a period equal to $9T$ seconds, and its spectrum must therefore be a line spectrum with a fundamental frequency of $2\pi/9T$ radians/second. Suppose we now evaluate its mean (that is, zero frequency) term and successive harmonics. The fourth harmonic will have a frequency $\frac{8}{9}(\pi/T)$ and the fifth harmonic $\frac{10}{9}(\pi/T)$: hence the above argument, which suggests that it is only necessary to evaluate the spectrum up to the frequency $\omega = \pi/T$, also suggests that in this case it would be pointless to evaluate harmonics above the fourth. Indeed, if the fifth were evaluated, its coefficient would be found equal to the complex conjugate of that of the fourth harmonic; the sixth would be similarly related to the third, and so on. Therefore the mean term and first four harmonics of the repetitive version of the signal shown in figure 4.5(b) are sufficient to specify it completely. Of course, the mere repetition of the original nine sample values of figure 4.5(a) has certainly not increased our

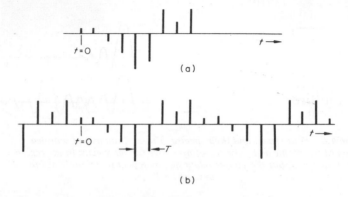

Figure 4.5   *(a) A signal consisting of nine sample values, and (b) its repetitive version*

information about the signal in any way, and we must therefore expect that our frequency-domain description in terms of a mean value and limited set of harmonics is just as valid for the original 9-sample signal.

It is perhaps not surprising that a signal of length 9 samples may be described completely by a zero-frequency term and four harmonics: each harmonic must be specified by two numbers (amplitude of sine and cosine components, or magnitude and phase terms), so that, in effect, such a frequency-domain description involves $[1 + (4 \times 2)] = 9$ separate numbers. Since time and frequency descriptions of a signal are essentially equivalent, we might expect the spectrum of a 9-sample signal to be definable by 9 numbers. In general, if we have a sampled-data signal of length N sample values, it is sufficient to evaluate its spectrum at $N/2$ spot frequencies in the range between $\omega = 0$ and $\omega = \pi/T$. This rather intuitive discussion will be given a more rigorous basis in chapter 8.

4.3.2 The fast Fourier transform

In order to find the spectrum of a sampled-data signal such as that already shown in figure 4.3, it is necessary to compute values of a function of the form

$$G(j\omega) = x_0 + x_1 e^{-j\omega T} + x_2 e^{-j\omega 2T} + x_3 e^{-j\omega 3T} + \ldots$$
$$= x_0 + x_1 \cos \omega T + x_2 \cos 2\omega T + x_3 \cos 3\omega T + \ldots$$
$$-j(x_1 \sin \omega T + x_2 \sin 2\omega T + x_3 \sin 3\omega T + \ldots)$$

The real and imaginary parts of this expression may be evaluated for a suitable set of values of ω, assuming the sampling interval T is known. As with continuous functions, an even sampled-data signal (that is, one which is symmetrical about $t = 0$) has a purely real spectrum and it is therefore only necessary to compute the cosines; conversely, if the time function is odd, it is only necessary to compute the sines.

It is interesting to consider the number of computer operations needed to evaluate the DFT of a signal consisting of N samples. As we have already seen, a signal of this length is defined by $N/2$ harmonics, and in the general case each is represented by both a cosine and a sine component. Evaluation of any one term involves multiplying each signal sample value by a factor of the form $\cos n\omega T$ or $\sin n\omega T$ and some N^2 products are therefore involved in calculating the spectrum of an N-sample signal. (There are also a comparable number of additions to be performed, but these are normally very much faster on a digital computer.) Therefore the time taken to calculate the Fourier transform is roughly proportional to the square of the number of sample values.

However, a more careful investigation shows that many of the multiplication operations involved are repeated, and the fast Fourier transform (FFT) aims to eliminate such repetitions as much as possible. This form of the DFT is therefore a recipe, or 'algorithm', which is designed to increase the efficiency of calculation. It many be shown[20] that the computation time for an N-sample transform is approximately proportional to $N \log_2 N$ when the FFT is used, as opposed to N^2 when no special precautions are taken to eliminate redundant calculations. Fast Fourier algorithms become more and more attractive as the number of signal samples increases, and are generally most efficient when this number is an integer power of 2 (say 1024 or 2048). With signals of this sort of length, reduction of computing time by a factor of 50 or 100 is commonly achieved.

In rather more detail, computation of the DFT involves estimation of $N/2$ discrete harmonics in the frequency range $0 < \omega < \pi/T$, which gives a spacing in ω between successive harmonics of $2\pi/NT$: the frequency of the kth harmonic is therefore $2\pi k/NT$. Denoting the value of $G(j\omega)$ corresponding to this spot frequency as $G(j\omega)_k$ we may write

$$G(j\omega)_k = x_0 + x_1 . \exp\{-j(2\pi k)/N\} + x_2 . \exp\{-j(2\pi k . 2)/N\}$$
$$+ x_3 . \exp\{-j(2\pi k . 3)/N\} + \ldots x_{N-1} . \exp[-j\{2\pi k(N-1)\}/N]$$

$$= \sum_{n=0}^{N-1} x_n \exp\{-j(2\pi kn)/N\}$$

where the capital sigma represents a summation of N terms. If the DFT is evaluated by using this formula as it stands, it is clear that the entire set of sample values, $x_0, x_1, \ldots, x_{N-1}$ is used to calculate each harmonic term $G(j\omega)_k$. Due to the periodic nature of $\exp\{-j(2\pi kn)/N\}$, this means that the same product $x_k . \exp\{-j(2\pi kn)/N\}$ is formed many times for different combinations of n and k. The FFT aims to minimise this repeated calculation of products by dividing the total sample set into a number of sub-groups; the transforms of these sub-groups are computed and stored, and later combined to produce the DFT. Division into sub-groups is most simple and efficient when the total number of samples is an integer power of 2.

4.4   The z-transform

4.4.1 Introductory

Although it is quite possible to use the Fourier or Laplace transforms to describe the frequency-domain properties of a sampled-data signal, there is another transform which is tailor-made for the purpose. The z-transform[9,10] provides not only a useful shorthand notation for the Fourier transform of such a signal, but also a very convenient method of defining the signal by a set of poles and zeros.

The z-transform of a sampled-data signal may be simply defined using its Laplace transform as a starting point. In section 4.3.1 it was shown that the Laplace transform of a signal such as that of figure 4.3 is given by

$$G(s) = x_0 + x_1 e^{-sT} + x_2 e^{-s2T} + x_3 e^{-s3T} + \ldots$$

We now define the new variable $z = e^{sT}$ and denote the z-transform of the signal by $G(z)$. Hence

$$G(z) = x_0 + x_1 z^{-1} + x_2 z^{-2} + x_3 z^{-3} + \ldots$$

z is a new frequency variable which generally has both real and imaginary parts. Thus if $s = \sigma + j\omega$

$$z = e^{sT} = e^{\sigma T}.e^{j\omega T} = e^{\sigma T} \cos \omega T + j e^{\sigma T} \sin \omega T$$

We have already seen that the term $e^{-j\omega T}$ implies a phase shift proportional to frequency and hence a constant time delay of T seconds. Conversely, the term $e^{j\omega T}$ represents a time advance of T seconds. This accounts for the common description of z as a 'shift operator', denoting a time advance equal to the sampling interval T.

It is clear from the above definition that the z-transform of a sampled signal is a power series in (z^{-1}) in which the coefficients of the various terms are equal to the sample values; when a transform is expressed in this form, it is possible to regenerate the time function merely by inspection. Suppose, for example, we have the z-transform

$$G(z) = \frac{z}{(z - \alpha)} = \frac{1}{(1 - \alpha z^{-1})}$$

We may rewrite this as a power series in z^{-1}

$$G(z) = 1 + \alpha z^{-1} + \alpha^2 z^{-2} + \alpha^3 z^{-3} + \alpha^4 z^{-4} + \ldots$$

and the corresponding time function is shown in figure 4.6 for two values of the constant α. These results demonstrate how a simple term such as $(z - \alpha)$ in the

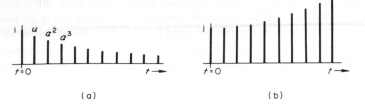

Figure 4.6   *Time function corresponding to the z-transform G(z) = z /(z − α)*
with (a) α < 1, and (b) α > 1

denominator of $G(z)$ represents a time function having an infinite set of sample values.

4.4.2 z-Plane poles and zeros

We have seen that the frequency spectrum of any sampled-data signal is repetitive in form with a period equal to $2\pi/T$ radians/second. Since a signal may be described by a set of s-plane poles and zeros whose positions bear a direct relationship to the signal spectrum, it must be expected that the s-plane pole–zero pattern of a sampled-data signal will also be repetitive in form. It is indeed possible to show[27] that the sampled version of a continuous signal has the same s-plane pole-zero configuration as the original, except that all poles and zeros are repeated indefinitely at intervals of $2\pi/T$ in the direction of the imaginary axis.

One of the advantages of using the s-plane pole–zero description of a continuous signal is that it allows the spectrum to be visualised quite easily, by considering the lengths and phases of vectors drawn from the various poles and zeros to a succession of points on the the imaginary axis (see section 3.4.3). In the case of a sampled-data signal, however, this technique becomes almost valueless, since there is an infinite set of poles and zeros to be taken into account. This difficulty emphasises a further advantage of the z-transform, namely that it allows the representation of a sampled-data signal by a finite set of poles and zeros.

In order to gain some insight into the relationships between s-plane and z-plane poles and zeros, it is useful to investigate what happens to the complex variable z when s takes on certain typical values. This process is generally referred to as 'mapping' the s-plane into the z-plane. Suppose, for example, s is purely imaginary. Then

$$s = j\omega \qquad \text{and} \qquad z = e^{sT} = e^{j\omega T} = \cos \omega T + j \sin \omega T$$

If we now consider values of z to be plotted on a z-plane Argand diagram, the locus of z will trace out a circle of unit radius as ω varies, starting on the real axis when $\omega = 0$ and repeating its trajectory at intervals in ω of $2\pi/T$. Next let $s = \sigma$, where σ is real, giving $z = e^{\sigma T}$. If σ is positive, z is a real positive number greater than 1; if σ is negative, z is a real positive number less than 1. Suppose finally that $s = \sigma + j\omega$, giving $z = e^{\sigma T} e^{j\omega T}$. This value of z is represented by a vector of length $e^{\sigma T}$ which makes an angle ωT radians with the positive real axis. Whenever σ is negative, a point inside the unit circle in the z-plane is specified which corresponds to a point in the left-hand half of the s-plane (that is, to the left of the imaginary axis). Therefore the complete left-hand half of the s-plane maps into the area inside the unit circle in the z-plane. These equivalent locations in the two planes are illustrated in figure 4.7.

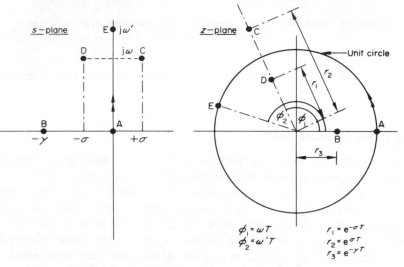

Figure 4.7   *Equivalent locations in the s and z planes*

Actually, this is not quite the whole story. Suppose we have a sampled-data signal described by a single z-plane pole at the point marked B in the right-hand part of figure 4.7. Its z-transform is then $G(z) = 1/(z - r_3)$, and its frequency spectrum is found by substituting $j\omega$ for s, giving

$$G(j\omega) = \frac{1}{(e^{j\omega T} - r_3)}$$

The denominator may be represented by a vector drawn from the point B to a point on the unit circle. Every time ω changes by $2\pi/T$ radians/second, the complete unit circle is traced out and the changes in length and phase of this vector—and therefore in the frequency spectrum—are repeated (the spectrum of a sampled-data signal may be inferred by considering vectors drawn from the poles and zeros to a point on the unit circle in a way analogous to that described for the s-plane in section 3.4.3). Hence a single z-plane pole at a point such as B is equivalent to an infinite repetitive set of s-plane poles, adjacent members of the set being separated by an interval of $2\pi/T$ in the direction of the imaginary axis. The s-plane point B in the left-hand half of figure 4.7 is merely one of this infinite set. This explains why the pole-zero description of a sampled-data signal is only economic if the z-plane is used.

The time functions corresponding to a few z-transforms will now be evaluated in order to give some familiarity with the method. Consider first the function

$$G_1(z) = (1 - z^{-8}) = \frac{z^8 - 1}{z^8}$$

There are eight zeros given by the roots of the equation $z^8 = 1$, and an eight-order pole at $z = 0$. Two of the zeros are clearly $z = 1$ and $z = -1$, and the other six are distributed around the unit circle in the z-plane as shown in figure 4.8. The

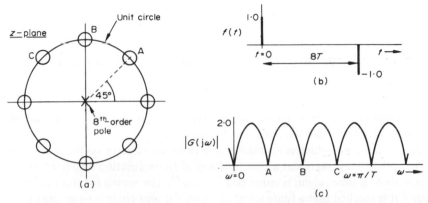

Figure 4.8　*(a) A z-plane pole–zero configuration, (b) the corresponding time function, and (c) the magnitude of its spectrum*

figure also shows the corresponding time function and its spectrum, the latter having three null points between $\omega = 0$ and $\omega = \pi/T$ corresponding to the three zeros labelled A, B, and C. If the zero at $z = 1$ is now removed, we have the new z-transform

$$G_2(z) = \frac{1}{z^8}\left[\left(z - \frac{1}{\sqrt{2}} + j\frac{1}{\sqrt{2}}\right)\left(z - \frac{1}{\sqrt{2}} - j\frac{1}{\sqrt{2}}\right)\left(z + \frac{1}{\sqrt{2}} + j\frac{1}{\sqrt{2}}\right)\right.$$
$$\left. \times \left(z + \frac{1}{\sqrt{2}} - j\frac{1}{\sqrt{2}}\right)(z + j)(z - j)(z + 1)\right]$$

which multiplies out to give

$$G_2(z) = z^{-1} + z^{-2} + z^{-3} + z^{-4} + z^{-5} + z^{-6} + z^{-7} + z^{-8}$$

The corresponding time function consists of a set of eight unit-height samples and is shown in figure 4.9. The spectrum differs from the first example principally by having a finite zero-frequency, or mean, term. It is interesting to note that this latter zero pattern may also be achieved by cancelling the zero at $z = 1$ by a coincident pole. Hence $G_2(z)$ may also be written in the form

$$G_2(z) = \frac{z^8 - 1}{z^8(z - 1)}$$

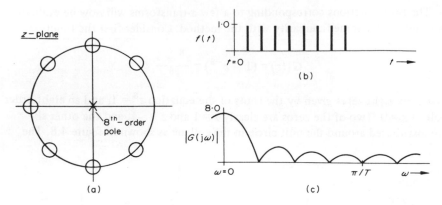

Figure 4.9   *The effects of removing one of the z-plane zeros of figure 4.8*

emphasising that a given z-transform may often be expressed more neatly as a set of poles and zeros rather than as a set of zeros alone.

Any z-transform which is expressible as a finite set of zeros on the unit circle of the z-plane must represent a time function of finite duration which, apart from a mere shift of time origin, is either odd or even[28]. The reason for this becomes clear if it is recalled that a finite set of zeros on the unit circle is equivalent to an infinite repetitive set on the imaginary axis in the s-plane. Any vector drawn from one such s-plane zero to a point on the imaginary axis must always have a phase angle of $\pm 90°$ ($\pi/2$ radians) and hence the total phase angle due to all such zero vectors must always be an integer multiple of $90°$. Thus the spectrum is always either purely real or purely imaginary, representing a time function composed of either cosines or sines. It is interesting to note that the time function of figure 4.8 is odd, and that of figure 4.9 is even, apart from a shift in time origin.

A slight variation on the theme of possible alternative representations of a signal by poles and zeros is provided by the sampled-data signal of figure 4.10(a), which is a truncated version of the 'infinite' decaying exponential waveform of figure 4.10(b). Since the truncated signal may be formed by subtracting the

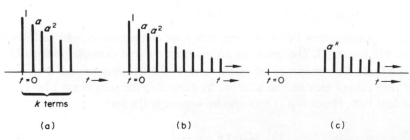

Figure 4.10   *Truncated and 'infinite' versions of a sampled exponential signal*

waveform of figure 4.10(c) from that of 4.10(b), its z-transform is written by inspection as

$$G_3(z) = \frac{z}{(z - \alpha)} - \alpha^k z^{-k} \frac{z}{(z - \alpha)}$$

$$= \frac{z}{(z - \alpha)}(1 - \alpha^k z^{-k})$$

It therefore has a set of k zeros equally spaced around a circle of radius α, except that the one at $z = \alpha$ is cancelled by a coincident pole, as shown in figure 4.11 for

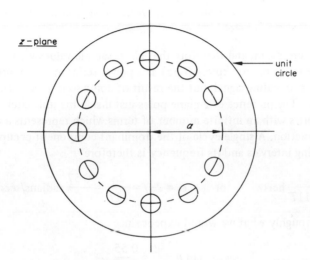

Figure 4.11   *z-plane zeros of a sampled exponential signal truncated to 12 terms*

the case when $k = 12$. It is interesting to note that if α and k are chosen so that all terms in the waveform of figure 4.10(c) are negligible, then the infinite and truncated version of the signal are effectively the same. And since the infinite version may be represented by a single pole at $z = \alpha$, we conclude that a sufficiently large group of zeros on a circle of radius α is equivalent to a single pole at 'the same radius.

Consider next the z-transform

$$G_4(z) = \frac{z(z - 1)}{(z - \alpha + j\beta)(z - \alpha - j\beta)} = \frac{z(z - 1)}{z^2 - 2\alpha z + (\alpha^2 + \beta^2)}$$

which has zeros at $z = 0$ and $z = 1$, and a complex conjugate pole pair at $z = \alpha \pm j\beta$. To take a numerical example suppose that $\alpha = 0.81$ and $\beta = 0.55$ which gives the pole positions shown in figure 4.12(a). These poles are close to the unit circle and therefore represent strong frequency components in the region of $\omega T = \theta$, or $\omega = \theta/T$ radians/second. The zero at $z = 1$ implies that the corresponding time

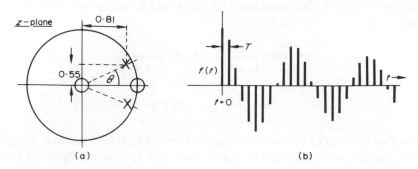

Figure 4.12   *(a) A pole–zero configuration which includes a complex conjugate pole pair just inside the unit circle, and (b) the corresponding time function*

function has zero mean, and therefore that all its sample values would sum to zero. It is not so simple to express $G_3(z)$ as a power series in z^{-1} in order to derive the corresponding signal, but the result of doing so is illustrated in figure 4.12(b). The presence of z-plane poles just inside the unit circle gives rise to a power series with an infinite number of terms which represents a decaying oscillatory function. A single cycle of the dominant component occupies about eleven sampling intervals and its frequency is therefore

$$f \approx \frac{1}{11T} \text{ hertz}, \quad \text{or} \quad \omega = 2\pi f = \frac{2\pi}{11T} = \frac{0\cdot57}{T} \text{ radians/second}$$

This value is roughly what we would expect since

$$\tan \theta = \frac{\beta}{\alpha} = \frac{0\cdot55}{0\cdot81}$$

and hence $\theta = 34°$, which is very close to $0\cdot57$ radians. The final point to make is that poles or zeros at the z-plane origin correspond to a pure time-shift and do not otherwise affect the form of the time function[28]. Thus the zero at $z = 0$ in the above example merely ensures that the first finite sample value occurs at $t = 0$; if the zero were not present, it would occur at $t = T$. Conversely, a single pole at the origin would correspond to a first sample value at $t = -T$. Generally speaking, the first nonzero sample occurs at $t = 0$ when the highest powers of z in the numerator and denominator polynomials of $G(z)$ are equal.

4.5   Discussion

This chapter has attempted to show how sampled-data signals may be fitted into the framework of frequency analysis concepts already developed. In discussing the spectra of sampled-data signals, the approach has been somewhat intuitive, and readers who wish for a fuller understanding of the relationships between the spectrum of a continuous signal and that of its sampled version will find such matters explored more carefully in chapter 8. So far we have merely assumed that a set of sample values representing some function of interest is to hand, and that we wish to investigate its frequency-domain properties without worrying unduly whether the samples form an adequate substitute for the underlying continuous function.

To say that the spectrum of a sampled-data signal is repetitive in form up to infinitely high frequencies may seem somewhat artificial. What this means, of course, is that if we wished to synthesise the sample set by summation of continuous sinusoidal waves (which is the basic concept of Fourier analysis) it would be necessary to include waves of infinitely high frequency; the reason for this is that we have chosen to represent sample values by infinitely narrow Dirac pulses. In a practical situation, it might be required to transmit a sequence of narrow electrical pulses representing signal samples through, say, a communications network. In this case the pulses would have some small but finite width, and their spectrum would contain less energy at very high frequencies than that of true Dirac pulses. But even if a practical sampled-data signal consisting of infinitely narrow pulses will never be encountered, the notion that the sampling process introduces components of far higher frequencies than those present in the underlying continuous waveform remains an important one.

The introduction to the z-transform given in this chapter is elementary, and is merely designed to show that the transform provides a more compact frequency-domain description of a sampled-data signal than the Fourier or Laplace transforms. This point is perhaps most clearly appreciated when the signal is described by a set of poles and zeros. It has been pointed out at the end of chapter 3 that the Laplace transform is useful for describing the frequency-domain properties of signal processing devices as well as those of signals, and this is no less true of the z-transform. We shall indeed see later that the z-transform is the natural tool for describing the performance of processing devices which operate on sampled-data signals, whether such devices are realised by specially-designed electronic circuits or, as is perhaps more likely, by the programming of a digital computer.

Problems

1. Derive expressions for the spectra of the sampled-data functions shown. Sketch the spectra and check that they are repetitive at intervals in ω of $2\pi/T$ radians/second.

2. Write a time-domain expression for the sampled-data signal of figure 4.5, given that successive sample values are as follows:

$$1, 2, -2, -5, -9, -7, 6, 3, 6.$$

Estimate the relative magnitudes of the spectral components at $\omega = 0$, $\omega = \pi/2T$ and $\omega = \pi/T$ radians/second.

3. Sketch the pole-zero configuration of the function

$$G(z) = \frac{(1 - z^4)}{(1 + z^2)z^2}$$

Hence write $G(z)$ in a form which specifies only z-plane zeros, and check your result by dividing the numerator of the above expression by its denominator.

4. Show that a pair of zeros at $z = \alpha$ and $z = 1/\alpha$ correspond to a (repeating) pair of s-plane zeros having equal but opposite real parts. Hence satisfy yourself that the corresponding time function must be symmetrical in form, and evaluate it.

5. A sampled-data signal has a sampling interval of 0·1 second and its z-transform has poles at $z = -0·4 \pm j\,0·90$. Assuming that there are no zeros close to these poles, estimate the frequency of the dominant component in the signal, and sketch the time function. What would be the effect if the poles were marginally outside the unit circle, rather than just within it?

5

Random Signals

5.1 Introduction

So far we have dealt with continuous and sampled signals having defined waveshapes. Such signals are described as 'deterministic', and the frequency spectra we calculate for them specify the magnitudes and relative phases of the sinusoidal waves which, if added together, would exactly resynthesise the original waveforms. By contrast, the value of a random waveform or signal is not specified at every instant of time, nor is it possible to predict its future with certainty on the basis of its past history. The reason for this is generally that we have insufficient understanding of the physical process producing the random signal; on other occasions, the understanding is present, but the effort involved in predicting the signal (or in describing its waveshape by a precise analytic function) is too great to be worth while. In such cases it is usual to evaluate some average properties of the signal which describe it adequately for the task in hand.

A common initial reaction is that a random signal with ill-defined properties can have little place in any scientific theory of signal analysis, but the opposite is in fact nearer the truth. Suppose, for example, it is desired to send a message along a telegraph link. It is almost valueless to send a known, deterministic, message since the person at the receiving end learns so little by its reception. As a simple example, it would be pointless to transmit a continuous sinusoidal tone, since once its amplitude, frequency and phase characteristics have been determined by the receiver, no further information is conveyed. But if the tone is switched on and off as in a Morse-code message, the receiver does not know whether a 'dot' or 'dash' is to be sent next, and it is this very randomness or uncertainty about the future of the signal which allows useful information to be conveyed. On the other hand it is quite possible to say something about the average properties of Morse-code messages, and such knowledge might be very useful to the designer of a telegraph system, even though he does not know what dot–dash sequence is to be sent on any particular occasion. Viewed in this light it is unsurprising that random signal theory has played a major role in the development of modern communications systems[2,4].

73

A signal may be random in a variety of ways. Perhaps the most common type of randomness is in the amplitude of a signal waveform, illustrated in figure 5.1(a). In this case future values of the signal cannot be predicted with certainty, even when its past history is taken into account. Figure 5.1(b) shows another common form of randomness, of the general type displayed by the Morse-code message just mentioned. Here the signal is always at one or other of two definite values but the transitions between these two levels occur at random instants. Signals in which the times of occurrence of some definite event or transition between states are

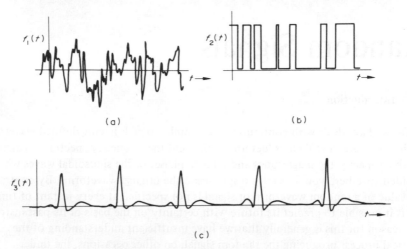

Figure 5.1   *Signals displaying (a) random amplitude, (b) random timing of transitions between fixed levels, and (c) random amplitude and timing components in an essentially repetitive waveform*

random arise in many diverse fields of study, such as queuing theory, nuclear particle physics and neurophysiology, as well as in electronic communications. Figure 5.1(c) shows a signal possessing both the common types of randomness so far mentioned, and represents an electrocardiogram (ECG), or recording of the electrical activity of the heart. The heartbeat is somewhat irregular, so that the timing of successive ECG complexes has a random component; furthermore, the waveshape itself is not exactly repetitive in form.

In practice a signal quite often contains both random and deterministic components. For example, the ECG complexes of figure 5.1(c) might usefully be considered to consist of a strictly repetitive signal plus small random amplitude and timing components. Very often such a random component is the result of recording or measurement errors, or arises at some point in the system which is outside the control of the experimenter. Random disturbances are widely encountered in electronic circuits, where they are referred to as electrical 'noise'. However, the methods of analysis of random waveforms described in this chapter apply whether a waveform represents a useful signal or an unwanted 'noise'. It is important to be able to describe a noise waveform quantitatively, not least so that the effects on it of signal processing devices may be assessed. And we shall see later (chapter 10) that a most important type of signal processing operation is one in which an attempt is made to extract or enhance a signal waveform in the presence of unwanted disturbances.

As already noted, the method used to describe random signals is to assess some average properties of interest. The branch of mathematics involved is statistics, which concerns itself with the quantitative properties of a population as a whole, rather than of its individual elements. In the present context, we may think of this population as being made up from a very large number of successive values of a random signal waveform. The other branch of mathematics of direct interest for random signal analysis is that of probability, which is closely related to statistics. Probability theory concerns itself with the likelihood of various possible outcomes of a random process or phenomenon, whereas statistics seeks to summarise the actual outcomes using average measures. In order to allow us to develop useful average measures for random signals, we first examine some of the basic notions of probability theory.

5.2 Elements of probability theory

5.2.1 The probability of an event

Suppose a die with six faces is thrown repeatedly. As the number of trials increases, and provided the die is fair, we expect that the number of times it lands on any one face to be close to one sixth of the total number of throws. If asked about the chance of the die landing on a particular face in the next trial, we would therefore assess it as one in six. Formally, if the trial is repeated N times and the event A occurs n times, the probability of event A is defined as

$$p(A) = \lim_{N \to \infty} \left(\frac{n}{N} \right)$$

This definition of probability as the relative frequency of the event (as the number of trials tends to infinity) has great intuitive appeal, although it gives rise to considerable difficulty in more rigorous mathematical treatments. The main reason for this is that any actual experiment necessarily involves a finite number of trials and may fail to approach the limit in a convincing way. Even in a simple coin-tossing experiment, a vast number of tosses may fail to yield the probabilities of 0·5 expected for both 'heads' and 'tails'. For this reason an alternative approach to the definition of probability, based upon a set of axioms, is sometimes adopted.[1] Whichever definition is used, however, it is clear that a probability of 1 denotes certainty and a probability of 0 implies that the event never occurs.

Next suppose that we define the event (A or B) as occurring whenever either A or B occurs. In a very large number N of trials, suppose A occurs n times and

B occurs m times. If A and B are mutually exclusive (so that they never occur together) the event (A or B) occurs $(n + m)$ times. Hence

$$p(\text{A or B}) = \lim_{N \to \infty} \left(\frac{n + m}{N} \right) = \lim_{N \to \infty} \left(\frac{n}{N} \right) + \lim_{N \to \infty} \left(\frac{m}{N} \right)$$

$$= p(\text{A}) + p(\text{B})$$

This is the basic additive law of probability, which may be extended to cover the case of an experiment or trial with any number of mutually-exclusive outcomes.

5.2.2 Joint and conditional probabilities

Suppose we now conduct an experiment with two sets of possible outcomes. Then the joint probability $p(\text{A and B})$ is the probability that the outcome A from one set occurs together with outcome B from the other set. For example, the experiment might consist of tossing two dice simultaneously, or of drawing two cards from a pack. Suppose that of N experiments performed n produce outcome A; of these, suppose that m also produce outcome B. Then m is the number of experiments which give rise to both A and B, so that

$$p(\text{A and B}) = \lim_{N \to \infty} \left(\frac{m}{N} \right)$$

$$= \lim_{N \to \infty} \left(\frac{n}{N} \right) \left(\frac{m}{n} \right)$$

Note that the limit of (n/N) as $N \to \infty$ is $p(\text{A})$, and that the limit of (m/n) is the probability that outcome B occurs given that outcome A has already occurred.

This latter probability is called the 'conditional probability of B given A' and will be denoted by the symbol $p(\text{B/A})$. Thus

$$p(\text{A and B}) = p(\text{A}) \cdot p(\text{B/A})$$

By similar arguments, it may be shown that

$$p(\text{A and B}) = p(\text{B}) \cdot p(\text{A/B})$$

and hence

$$p(\text{B}) \cdot p(\text{A/B}) = p(\text{A}) \cdot p(\text{B/A})$$

or

$$p(\text{A/B}) = \frac{p(\text{A}) \cdot p(\text{B/A})}{p(\text{B})}$$

This result, known as Bayes' rule, relates the conditional probability of A given B to that of B given A.

It is obvious that if outcome B is completely unaffected by outcome A, then the conditional probability of B given A will be just the same as the probability of B alone, hence

$$p(B/A) = p(B)$$

and

$$p(A \text{ and } B) = p(A) \cdot p(B)$$

In this case the outcomes A and B are said to be statistically independent.

To illustrate these results, suppose we have a box containing 3 red balls and 5 blue balls. The first experiment consists of taking one ball from the box at random, replacing it and then taking another, and we are interested in the probability that the first ball withdrawn is red (outcome A) and the second one blue (outcome B). In this case, the result of the first part of the experiment in no way affects that of the second, since the first ball is replaced. Hence the probability of the joint event (red followed by blue) is

$$p(A \text{ and } B) = p(A) \cdot p(B) = \tfrac{3}{8} \cdot \tfrac{5}{8} = \tfrac{15}{64}$$

We now repeat the experiment without replacing the first ball before taking out the second, so that the two parts of the experiment are no longer statistically independent. If a red ball is withdrawn first, then 5 blue and 2 red balls remain, giving the conditional probability

$$p(B/A) = \tfrac{5}{7}$$

Therefore the probability of the joint event is now changed to

$$p(A \text{ and } B) = p(A) \cdot p(B/A) = \tfrac{3}{8} \cdot \tfrac{5}{7} = \tfrac{15}{56}$$

This brief discussion of joint and conditional probabilities, which has been applied to the case of an experiment with two sets of possible outcomes, may be extended to cover any number of such sets. Its relevance to signals and signal analysis may be illustrated by reference to figure 5.2(a), which shows a portion of a random sampled-data signal in which each sample value takes on one of six possible levels. This is analogous to the die with six faces, and we could estimate the probability of the next sample taking on any particular value by analysing a sufficiently long portion of the signal's past history (note that, unlike the case of the fair die, the six probabilities may well not be equal). Thus any one sample value is thought of as the result of a trial which has, in this case, six possible outcomes. A rather more elaborate description of the signal could be obtained by asking whether a particular value of the signal tends to influence following ones. For example, does the occurrence of a sample 3 mean that a 4 or 2 (or any other value) is more likely to follow than would be suggested by the simple probabilities already derived? This is tantamount to asking whether a particular value is constrained by those preceding or following it and could be answered by assessing suitable conditional probabilities. Such matters will be investigated in subsequent sections where so-called correlation functions are discussed.

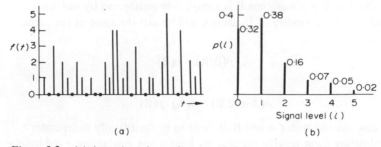

Figure 5.2   *(a) A portion of a random sampled-data signal with six discrete levels and (b) the probabilities associated with each level*

So far we have considered experiments having a finite number of possible outcomes, and the analogous situation of a random signal which takes on a finite number of levels. In the case of a signal such as that of figure 5.2(a), analysis of a very long portion of the signal would allow us to estimate the probabilities associated with the six possible signal levels; these could then be plotted on a graph as in figure 5.2(b). Commonly met in practice, however, are signals which display a continuous range of amplitude levels. The probability of finding such a signal at a particular level then becomes vanishingly small and we are forced to use a continuous probability variable, the probability density function, to describe it.

5.2.3 The probability density function

The continuous random signal shown in figure 5.3 takes on an infinite set of amplitude values, and the probability of its assuming some particular value such as y therefore becomes vanishingly small. In this case it is only sensible to talk about the probability that the signal occupies a small range of values such as that between y and $(y + \delta y)$. This is simply equal to the fraction of the total time spent by the signal in this range, and may be estimated by summing all time intervals such as $\delta t_1, \delta t_2, \delta t_3 \ldots$ over a long length of record (say T_0 seconds) and then dividing the result by T_0. Denoting this probability by q, we have

$$q = \lim_{T_0 \to \infty} \left(\frac{\delta t_1 + \delta t_2 + \delta t_3 + \ldots}{T_0} \right)$$

Now the value of q clearly depends on the size of the interval δy: indeed, as δy tends to zero, so does q. However, the quotient $(q/\delta y)$ tends to a constant value

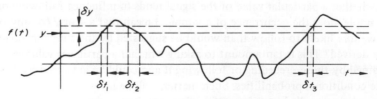

Figure 5.3   *A continuous random signal*

as $\delta y \to 0$, and this value is called the probability density $p(y)$. $p(y)$ gives us a probability measure which is independent of the precise value of δy chosen: the actual numerical probability that the signal falls in the range y to $(y + \delta y)$ is found by multiplying $p(y)$ by the interval size δy; hence

$$p(y) = (q/\delta y) \qquad \text{or} \qquad q = p(y) . \delta y$$

It was pointed out in section 3.3.1 that the spectrum $G(j\omega)$ of an aperiodic time function should be thought of as a density function, because it is only sensible to talk about spectral energy in some narrow band of frequencies, rather than at any particular spot frequency. The probability density function of a random signal involves a similar concept: it is not sensible to talk about the probability that a continuous signal adopts a particular value y, but only that it falls within some narrow range of values between y and $(y + \delta y)$.

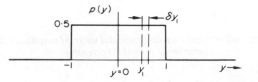

Figure 5.4   *A rectangular probability density function*

To take a simple example, suppose we have a continuous random signal $f(t)$ such that $-1 < f(t) < 1$ for all t, with no bias towards any particular range within these limits. In other words $p(y)$ is a 'rectangular' or 'even' distribution having a constant value in the range $-1 < y < 1$, as shown in figure 5.4. The actual numerical value of $p(y)$ may be found by considering a small range of values such as y_1 to $(y_1 + \delta y_1)$. The probability of finding the signal in this range is

$$\frac{\delta y_1}{1 - (-1)} = \frac{\delta y_1}{2} = p(y_1) . \delta y_1$$

$$\therefore \quad p(y_1) = 0.5.$$

Therefore in this particular case the probability density has a constant value of 0.5 in the range $-1 < y < 1$. Generalising to any form of function $p(y)$, the probability of finding the signal somewhere in the range $-\infty < y < \infty$ must be unity, and hence

$$\int_{-\infty}^{\infty} p(y) . dy = 1$$

In other words, any probability density function must have unit area.

5.3   Amplitude distributions and moments

We have already seen two examples (figures 5.2(b) and 5.4) of functions which
describe the probability that a random signal takes on a particular value, or falls
within a certain narrow range of values. Such functions are often referred to as
'amplitude distributions', and, like other long-term or average measures used to
describe random signals, they may also be applied to deterministic ones. For
example, the sinusoidal wave of figure 5.5(a) which spends a relatively large

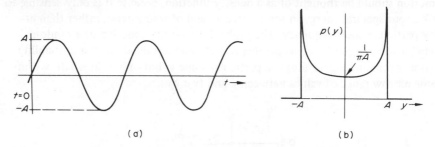

Figure 5.5   *(a) Part of a continuous sinusoidal wave of amplitude A, and (b) its
amplitude probability density*

proportion of its time near the values $\pm A$ but never exceeds them, has the
amplitude probability density curve shown in figure 5.5(b). On the whole, however,
such amplitude probability functions are not used for deterministic signals because
compact (and complete) analytic descriptions of the functions themselves are
normally available.

Another important point is that the amplitude distribution of a signal tells us
nothing about its detailed structure or spectrum, nor is it a unique property of the
signal. Figure 5.6 shows portions of one random and three deterministic waveforms
all having identical amplitude distributions but with very different structures and
frequency spectra.

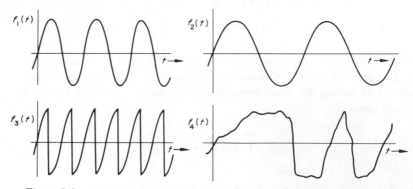

Figure 5.6   *Portions of four waveforms with identical amplitude distributions*

It is sometimes of interest to know the probability that a signal has a value below a certain level a. Denoting this probability by $P(a)$, we have

$$P(a) = \int_{-\infty}^{a} p(y) \cdot dy$$

P is normally referred to as the 'cumulative' distribution function. Alternatively, if the probability of finding the signal above some level b is required, it is only necessary to change the limits of integration to b and $+\infty$.

A further very important set of properties of a random (or deterministic) signal, known as the central moments, may be evaluated if the amplitude distribution function is known. The first central moment, more widely known as the average or mean value, and the second central moment or variance, are the most important, although higher order ones are sometimes also of interest[29]. Consider first the case of a random signal of the general type already shown in figure 5.2, which can assume n discrete values $y_1, y_2, y_3 \ldots y_n$. If we consider a very large number ($N \to \infty$) of sample values, the value y_1 is expected to occur Np_1 times, where p_1 is the probability associated with it; value y_2 is expected to occur Np_2 times, and so on. The average, or mean, of all the sample values is found by adding together all the sample values multiplied by the frequency with which they occur, and dividing by the total number of samples. Hence the average ($\bar{y}$) is given by

$$\bar{y} = \lim_{N \to \infty} \frac{1}{N} (Np_1 y_1 + Np_2 y_2 + Np_3 y_3 + \ldots Np_n y_n)$$

$$= \sum_{m=1}^{n} p_m y_m$$

where the capital sigma represents a summation of n terms. This average, or mean, value is also called the 'expected' value of y and is often written as $E(y)$ or $\langle y \rangle$. When the random signal is continuous and can take up an infinite set of values, the summation is replaced by an integration, and the discrete probability function is replaced by a probability density. In this case the average value is

$$\bar{y} = \int_{-\infty}^{\infty} y \cdot p(y) \cdot dy$$

The second central moment, or variance, is generally given the symbol σ^2, where σ is referred to as the 'standard deviation'. The variance is defined as the average or expected value of the square of the function's departure from its mean, and is therefore a measure of signal fluctuation. By arguments similar to those used when defining the mean, the second central moment is given by

$$\sigma^2 = \overline{(y - \bar{y})^2} = \sum_{m=1}^{n} (y_m - \bar{y})^2 \cdot p_m$$

for a discrete random signal having n possible levels, and as

$$\sigma^2 = \overline{(y - \bar{y})^2} = \int_{-\infty}^{\infty} (y - \bar{y})^2 . p(y) . dy$$

for a continuous one.

As an example, let us now evaluate first and second central moments for the random signal of figure 5.2. Using the probabilities for the six possible signal levels shown in figure 5.2(b), the first central moment or mean is given by

$$\bar{y} = (0\cdot32 \times 0) + (0\cdot38 \times 1) + (0\cdot16 \times 2) + (0\cdot07 \times 3)$$
$$+ (0\cdot05 \times 4) + (0\cdot02 \times 5)$$
$$= 1\cdot21$$

To find the second central moment, or variance, we now subtract this mean of $1\cdot21$ from each of the possible signal values, square the result, and multiply by the appropriate probability, giving

$$\overline{(y - \bar{y})^2} = (0 - 1\cdot21)^2 . 0\cdot32 + (1 - 1\cdot21)^2 . 0\cdot38 + (2 - 1\cdot21)^2 . 0\cdot16$$
$$+ (3 - 1\cdot21)^2 . 0\cdot07 + (4 - 1\cdot21)^2 . 0\cdot05 + (5 - 1\cdot21)^2 . 0\cdot02$$
$$= 1\cdot48$$

By cubing rather than squaring the bracketed terms we could derive the third central moment, and so on, although moments of higher order than the second are not often used. It is however worth noting two things: firstly, if the probability distribution is symmetrical in form, the third and higher odd-order central moments must be zero, and therefore such moments give an indication of asymmetry or 'skew' in a distribution: and secondly, moments of higher order pay increasing attention to the extreme values of a signal.

A distinction must be made between the 'central moments' and the 'moments' of a function. The former are measures of fluctuation about the mean value, whereas the simple moments are calculated without taking the mean into account. Thus the second moment of a discrete random variable with n possible levels is simply

$$\overline{y^2} = \sum_{m=1}^{n} y_m^2 . p_m$$

Not surprisingly, the central moments and moments are closely related and knowledge of one set allows the other to be calculated[29]. In particular, the second moment is related to the second central moment and the mean as follows

$$\overline{y^2} = \sigma^2 + (\bar{y})^2$$

This relationship explains why the second central moment or variance is such an important measure. As has already been pointed out in section 2.4.2, the average squared value of a signal (its second moment) equals its average power, adopting the convention that the signal is considered to represent an electrical voltage across or current through a 1 ohm resistor. Now the square of the mean value represents the power in the zero frequency (d.c.) component of the signal, and therefore the second central moment (σ^2) must represent the power in all other frequency components. It is therefore widely referred to as the 'a.c. power'.

Before leaving the topic of moments, the idea of a 'moment generating function' will be briefly discussed. Suppose we have a random signal with a probability density function $p(y)$, and we wish to estimate the mean and higher-order moments. If we multiply $p(y)$ by the function e^{ay} and integrate between $y = \pm\infty$, we specify the integral

$$I = \int_{-\infty}^{\infty} p(y).e^{ay}.dy$$

Now e^{ay} may be expressed as the power series

$$e^{ay} = 1 + ay + \frac{(ay)^2}{2!} + \frac{(ay)^3}{3!} + \ldots$$

$$= 1 + ay + \frac{a^2}{2!}.y^2 + \frac{a^3}{3!}y^3 + \ldots$$

Hence

$$I = \int_{-\infty}^{\infty} p(y).dy + a \int_{-\infty}^{\infty} p(y).y.dy + \frac{a^2}{2!} \int_{-\infty}^{\infty} p(y).y^2.dy + \ldots.$$

The first term of this series is equal to 1, and the subsequent ones are merely the moments of the random signal, multiplied by a constant. Hence

$$I = 1 + a\bar{y} + \frac{a^2}{2!}\overline{y^2} + \frac{a^3}{3!}\overline{y^3} + \ldots$$

Thus if we can evaluate the integral I in a particular case and express the result as a power series in a, we are immediately in a position to write down the moments. For example, suppose $p(y)$ is of the rectangular form shown already in figure 5.4, so that

$$I = \int_{-\infty}^{\infty} p(y).e^{ay}.dy = \int_{-1}^{1} 0.5 \, e^{ay} \, dy = \frac{1}{2a}\left[e^{ay}\right]_{-1}^{1}$$

$$= \frac{1}{2a}(e^a - e^{-a}) = \frac{2}{2a}\left[a + \frac{a^3}{3!} + \frac{a^5}{5!} + \frac{a^7}{7!} + \ldots\right]$$

$$= 1 + \frac{a^2}{3!} + \frac{a^4}{5!} + \frac{a^6}{7!} + \ldots$$

We now equate the coefficients of equal powers of a, giving

$$\bar{y} = 0, \qquad \overline{y^2} = \tfrac{1}{3}, \qquad \overline{y^3} = 0, \qquad \overline{y^4} = \tfrac{1}{5}, \qquad \text{etc.}$$

The mean value of a signal with this amplitude distribution is therefore zero, its second moment or average power is 1/3, and so on. These results have been very simply derived using the auxiliary function e^{ay}, which is called the 'moment-generating function'. The observant reader will have noticed that the integral I bears a striking similarity to a Fourier transform expression, except that y replaces the normal symbol t and the symbol a takes the place of $(-j\omega)$. Indeed, transform operations are used in probability theory for applications other than the generation of moments, and the Fourier transform of a probability density function is sufficiently widely used to be given a special name, the 'characteristic function'. The benefits of using characteristic functions rather than the probability density functions themselves are described in any advanced book on probability and statistics[14,15].

5.4   The autocorrelation and power spectral density functions

5.4.1 The spectral properties of random signals

The probability functions which we have so far investigated provide no clues to the structure of a random signal in the time-domain, or to its frequency spectrum. At first it may not be obvious that the spectrum of a random signal may be discussed at all, since the signal is unpredictable and is not defined by an analytic function. However, figure 5.7, which shows portions of two random signals, is

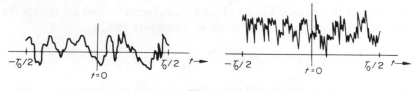

Figure 5.7

sufficient evidence that a spectral measure is likely to be very useful, assuming that one may be adequately defined. Suppose we treated these two portions as deterministic signals, by assuming them to have the waveforms shown in the interval $-T_0/2 < t < T_0/2$ and to be zero elsewhere, and then evaluated their frequency spectra. It is obvious that the two spectra would be quite different, if only because one signal has far higher frequencies present than the other and also possesses a substantial mean (zero frequency) component. If we now took a second portion of each random signal of the same duration T_0 and again evaluated the spectra, we would be bound to find them different from the first pair because of detailed differences in the waveforms. But the longer the duration T_0, the more confident would we be that each waveform portion was typical of its parent signal, and therefore that spectra calculated for different portions of a particular random signal would show broad similarities. In other words, we expect that some useful average measure of spectral components can be found, even if the spectrum of any finite portion of a waveform can never be expected to match it perfectly. The average measure most widely adopted is the so-called power spectrum, or its associated time-domain function, the autocorrelation function.

5.4.2 The autocorrelation function

The autocorrelation function (ACF) of a signal waveform is an average measure of its time-domain properties, and is therefore likely to be especially relevant when the signal is a random one. Furthermore, we shall see that the ACF is not only an interesting and valuable function in its own right, but that it also provides the key to a random signal's spectrum. Formally, the ACF is defined as

$$r_{xx}(\tau) = \lim_{T_0 \to \infty} \frac{1}{T_0} \int_{-T_0/2}^{T_0/2} f(t) \cdot f(t + \tau) \, . \, dt$$

It is therefore equal to the average product of the signal $f(t)$ and a time-shifted version of itself, and is a function of the imposed time-shift, τ. The above expression applies to the case of a continuous signal of infinite duration. If used for a signal of limited duration such as an isolated pulse, the average product over a very long interval T would tend to zero at all values of τ. In such cases it is therefore normal to use a modified version of the ACF, generally called the 'finite ACF', which is defined as

$$r'_{xx}(\tau) = \int_{-\infty}^{\infty} f_1(t) \cdot f_1(t + \tau) \, . \, dt$$

In the case of a sampled-data signal the product of the signal and its shifted version only has non-zero values when the shift is equal to a multiple of the sampling interval T, and the ACF is therefore defined as

$$r_{xx}(k) = \lim_{N \to \infty} \frac{1}{(2N + 1)} \sum_{m=-N}^{N} x_m \cdot x_{m+k}$$

where x_m and x_{m+k} represent two sample values separated by kT seconds, and the summation has $(2N + 1)$ terms with the integer parameter m taken between $-N$ and $+N$. In the case of a signal of limited duration it is once again appropriate to use a 'finite' version of the above expression which takes either the average or just the sum of a number of terms corresponding to the available length of signal. It should be noted that the precise definition of the autocorrelation function for finite and infinite signals tends to vary somewhat from text to text; the important thing to remember, however, is that they are all measures of the average product of a signal and its time-shifted version. (The subscript 'xx' or '11' is widely used for the ACF to denote that a signal is being multiplied by a delayed version of itself. This distinguishes it from the closely related cross-correlation function, which we shall use in the next chapter to describe the effects of time-shifts imposed between two different signals, and which is generally given the subscript '12' or 'xy'.)

Like the average measures described in the previous section, the ACF may be applied to deterministic as well as to random signals, and some of its principal properties are most conveniently illustrated in this way. Suppose for example we have a signal $f(t)$ composed of two cosinusoidal waves of different frequency and phase angle. Thus

$$f(t) = A_1 \cos(\omega_1 t + \theta_1) + A_2 \cos(\omega_2 t + \theta_2)$$

the ACF is then given by

$$r_{11}(\tau) = \lim_{T_0 \to \infty} \frac{1}{T_0} \int_{-T_0/2}^{T_0/2} [A_1 \cos(\omega_1 t + \theta_1) + A_2 \cos(\omega_2 t + \theta_2)]$$

$$\times [A_1 \cos(\omega_1(t + \tau) + \theta_1) + A_2 \cos(\omega_2(t + \tau) + \theta_2)] . dt$$

The integrand may be expanded as follows

$$A_1^2 \cos(\omega_1 t + \theta_1) . \cos[\omega_1(t + \tau) + \theta_1]$$
$$+ A_2^2 \cos(\omega_2 t + \theta_2) . \cos[\omega_2(t + \tau) + \theta_2]$$
$$+ A_1 A_2 \cos(\omega_1 t + \theta_1) . \cos[\omega_2(t + \tau) + \theta_2]$$
$$+ A_1 A_2 \cos[\omega_1(t + \tau) + \theta_1] . \cos(\omega_2 t + \theta_2)$$

Each of the last two terms represents a wave of frequency ω_1 multiplied by another of frequency ω_2, which averages out to zero over a very long interval T_0. The reason for this is the orthogonality of sine and cosine waves of different frequencies discussed already in section 2.3.2. The contribution to the integral of the other two terms may be found by using the trigonometric identity

$$\cos A . \cos B = \tfrac{1}{2}[\cos(A + B) + \cos(A - B)]$$

Therefore

$$A_1^2 \cos(\omega_1 t + \theta_1) . \cos[\omega_1(t + \tau) + \theta_1] = \frac{A_1^2}{2} [\cos(2\omega_1 t + \omega_1 \tau + 2\theta_1)$$

$$+ \cos(-\omega_1 \tau)]$$

and

$$A_2^2 \cos(\omega_2 t + \theta_2) . \cos[\omega_2(t + \tau) + \theta_2] = \frac{A_2^2}{2} [\cos(2\omega_2 t + \omega_2 \tau + 2\theta_2)$$

$$+ \cos(-\omega_2 \tau)]$$

Now a term such as $\cos(2\omega_1 t + \omega_1 \tau + 2\theta_1)$ represents a simple cosine wave of frequency $2\omega_1$ with a phase angle depending upon the value of τ considered; its average value over a very long time interval is therefore also zero. On the other hand, the term $\cos(-\omega_1 \tau) = \cos \omega_1 \tau$ is just a constant, since we are integrating with respect to t. Hence only the terms $\cos \omega_1 \tau$ and $\cos \omega_2 \tau$ produce contributions to the integral and we are left with

$$r_{11}(\tau) = \lim_{T_0 \to \infty} \frac{1}{T_0} \int_{-T_0/2}^{T_0/2} \left(\frac{A_1^2}{2} \cos \omega_1 \tau + \frac{A_2^2}{2} \cos \omega_2 \tau \right) dt$$

$$= \lim_{T_0 \to \infty} \frac{1}{T_0} \left(\frac{A_1^2}{2} \cos \omega_1 \tau + \frac{A_2^2}{2} \cos \omega_2 \tau \right) \cdot \left[t \right]_{-T_0/2}^{T_0/2}$$

$$= \frac{A_1^2}{2} \cos \omega_1 \tau + \frac{A_2^2}{2} \cos \omega_2 \tau$$

Therefore each of the frequency components in the signal $f(t)$ gives rise to a term in the autocorrelation function having the same period in the time-shift variable τ as the original component has in the time variable t, and an amplitude equal to half the squared value of the original. This result is illustrated by figure 5.8. The

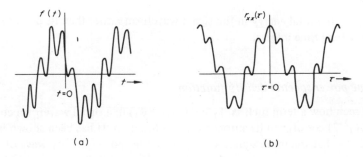

Figure 5.8 *(a) A signal with two discrete frequency components, and (b) its autocorrelation function*

phase shifts θ_1 and θ_2 of the component waves of $f(t)$ do not figure in the ACF, and hence all information about relative phases has been lost. Although the above ACF has been obtained for a function $f(t)$ consisting of just two component frequencies, the orthogonal properties of sine and cosine functions mean that it may be extended to the case of a signal containing any number of components. Regardless of its phase, each and every component gives rise to a simple cosine term in the ACF.

Since any ACF is composed of cosines it must be an even function of τ. In other words the averaged product of a signal and its time-shifted version is the same whether the shift is a forward or backward one, a conclusion which is indeed quite simple to demonstrate. When $\tau = 0$, all the cosine functions are at their peak positive value and hence reinforce one another to give the largest possible value of the ACF. Whether this peak value is ever attained again at other values of τ depends upon whether or not the various components in the signal are harmonically related, but in any event it can never be exceeded. The peak value is given by

$$r_{xx}(\tau)\,|_{\tau=0} = r_{xx}(0) = \lim_{T_0 \to \infty} \frac{1}{T_0} \int_{-T_0/2}^{T_0/2} [f(t)]^2 \, . \, dt$$

which is simply the mean square value, or average power, of the signal. Thus $r_{xx}(0)$ is equal to the second moment of the signal which, as we have seen, may also be derived from its probability density function. On the other hand, the curve traced out by the ACF as τ varies gives information about the time-domain structure of a signal which is not contained in the probability density function. It should be noted that in the case of a signal of finite duration, for which we use the 'finite' version of the ACF, the value of the ACF relevant to $\tau = 0$ is given by

$$r_{xx}(0) = \int_{-\infty}^{\infty} [f(t)]^2 \, . \, dt$$

which equals the total energy in the signal waveform rather than its power averaged over a very long time interval.

5.4.3 The power spectral density function

We have seen how a term such as $A_1 \cos(\omega_1 t + \theta_1)$ in a signal waveform contributes a term $(A_1^2/2) \cos \omega_1 \tau$ to its autocorrelation function. It has been shown in section 2.4.2 that the mean square value, or average power, of any wave of sinusoidal form having an amplitude A_1 is equal to $A_1^2/2$. Therefore the amplitudes of the various cosine terms in the ACF merely indicate the average power of the corresponding spectral terms in the signal itself.

Just as a signal waveform may be described in terms of its frequency spectrum, so an autocorrelation function (which is a function of the time-shift variable τ) has a counterpart in the frequency domain. As the above argument shows, this counterpart will have a number of spectral lines representing the power in the various components, and it is therefore given the name 'power spectrum'. The relationships between the frequency components of a typical periodic signal and those of its ACF are illustrated in figure 5.9. In the case of an aperiodic signal with a continuous frequency spectrum, the frequency-domain counterpart of its ACF is also continuous and is known as the 'power spectral density'.

Separate provision must again be made for signals of limited duration, whose average power, measured over a very long time interval, tends to zero. In such cases we refer to the 'energy spectrum' rather than the power spectrum. Just as the power spectrum is the frequency domain counterpart of the normal version of the ACF, so the energy spectrum is the equivalent of the 'finite' version of the ACF, and describes the distribution of signal energy in the frequency domain.

The discussion which follows (both here and in subsequent sections) will be largely in terms of the power spectrum of a signal which continues for ever, but it may be assumed that the energy spectrum of a time-limited signal displays essentially similar properties. Both power and energy spectra will be denoted by the symbol $P_{xx}(\omega)$.

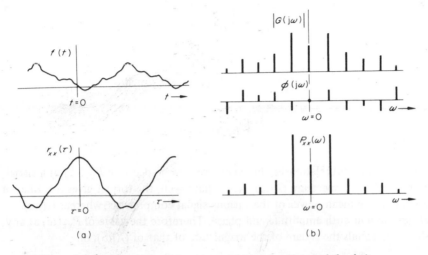

Figure 5.9 *(a) A signal and its autocorrelation function, and (b) their corresponding frequency-domain descriptions $G(j\omega)$ and $P_{xx}(\omega)$. $G(j\omega)$ involves both magnitude and phase terms, whereas $P_{xx}(\omega)$ is purely real*

We therefore see that the ACF and power spectrum, or power spectral density, are equivalent measures in the time and frequency domains. In other words, they are related to one another by the Fourier transform. This fact is formally expressed by the so-called Wiener–Khinchin relations[1]

$$P_{xx}(\omega) = \int_{-\infty}^{\infty} r_{xx}(\tau) . e^{-j\omega\tau} . d\tau$$

and

$$r_{xx}(\tau) = \frac{1}{2\pi} \int_{-\infty}^{\infty} P_{xx}(\omega) . e^{j\omega\tau} . d\tau$$

where $P_{xx}(\omega)$ is the power spectral density. These equations are identical in form to the Fourier transform pair which relate a signal $f(t)$ to its spectrum $G(j\omega)$ (see section 3.3.1), except that we are now working in terms of the delay variable τ rather than the time variable t. Furthermore, since the ACF is always an even function of τ, the power spectrum is always an even function of ω and is purely real. The imaginary parts of $e^{-j\omega\tau}$ and $e^{j\omega\tau}$ do not therefore contribute to their respective integrals and we may write

$$P_{xx}(\omega) = \int_{-\infty}^{\infty} r_{xx}(\tau) . \cos \omega\tau . d\tau$$

$$= 2 \int_{0}^{\infty} r_{xx}(\tau) . \cos \omega\tau . d\tau$$

and

$$r_{xx}(\tau) = \frac{1}{2\pi} \int_{-\infty}^{\infty} P_{xx}(\omega) . \cos \omega\tau . d\omega$$

$$= \frac{1}{\pi} \int_{0}^{\infty} P_{xx}(\omega) . \cos \omega\tau . d\omega$$

There is a useful relationship between the power spectrum $P_{xx}(\omega)$ of a signal and its frequency spectrum $G(j\omega)$. As we have seen, the former gives information only about the mean power of the various signal components, whereas $G(j\omega)$ defines them in both amplitude and phase. Therefore the value of $P_{xx}(\omega)$ at any value of ω equals the square of the magnitude of that of $G(j\omega)$, or

$$P_{xx}(\omega) = |G(j\omega)|^2$$

Furthermore if we write $G(j\omega)$ as $[a(\omega) + j . b(\omega)]$, then

$$|G(j\omega)| = \sqrt{[a^2(\omega) + b^2(\omega)]}$$

$$\therefore \qquad P_{xx}(\omega) = a^2(\omega) + b^2(\omega) = [a(\omega) + j.b(\omega)] [a(\omega) - j.b(\omega)]$$

$$= G(j\omega) . G^*(j\omega)$$

where the asterisk denotes the complex conjugate. $G^*(j\omega)$ has the same magnitude characteristic as $G(j\omega)$ but an equal and opposite phase characteristic; when the two are multiplied together their phase terms cancel.

5.4.4 Application to random signals

So far, the properties of the autocorrelation and power spectral density functions have been discussed by applying them to deterministic signals. Consider now a signal in which all frequencies between, say, $-\omega_0$ and $+\omega_0$ are equally represented in magnitude but have purely arbitrary relative phases. A typical portion of such a waveform and its power spectral density characteristic are illustrated in figure 5.10(a) and (b). It is clear that another waveform such as that of figure 5.10(c) in which the relative phases of the various terms have been rearranged will have the same power spectrum, since the latter measure takes no account of phase. We may therefore envisage a whole family of waveforms having quite random phase relationships but sharing the same power spectral density characteristic.

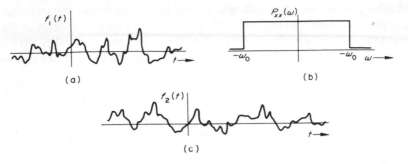

Figure 5.10 *(a) Typical portion of a random signal, and (b) its power spectral density which also applies to the second random signal shown in (c)*

The autocorrelation function corresponding to the power spectral density of figure 5.10(b) is of $(\sin x/x)$ form and is shown in figure 5.11. From this figure it is clear that the average product of the waveform and a shifted version of itself is relatively large when the shift is small. The reason for this is not hard to visualise: when the shift is small, large positive values of the wave tend to be multiplied by large positive values, and large negative values by large negative values, giving a large positive product. As the shift increases there is less and less tie-up between similar values of the wave and its shifted version, and the average product therefore tends to zero. The higher the frequencies present in the waveform, the faster it changes and the smaller the shift necessary to reduce the correlation to a very small value. This is shown by the fact that successive zero-crossings of the ACF occur at multiples of π/ω_0 seconds. As ω_0 increases the width of the main lobe of the $(\sin x/x)$ function reduces, and in the limit the ACF becomes a very narrow pulse centred on $\tau = 0$. It is interesting to note that the ACF is then similar in form to that of a Dirac pulse (see section 4.2), which shows that the basic difference between such a pulse and the waveforms shown in figure 5.10(a) and (c) lies in the latters' random phase relationships.

In section 5.2.2 the idea of a conditional probability was discussed, and it is worth noting that autocorrelation and conditional probability are closely related concepts. The value of an ACF at some time-shift τ is a measure of the average relationship between values of a signal separated by τ seconds. If, on average, a

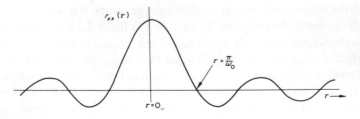

Figure 5.11 *Autocorrelation function corresponding to the power spectral density function of figure 5.10(b)*

signal value at some time t_1 is followed at time $(t_1 + \tau)$ by one of the same sign, the correlation function's value will be positive, and vice versa. The value of the ACF at time-shift τ is therefore a measure of the conditional probability that, given a positive (or negative) signal value at a particular instant, a value of like sign will occur τ seconds later.

It is important to realise that the power spectral density and autocorrelation functions so far illustrated are theoretical curves applying to a random signal of infinite extent, and that practical estimates derived from limited portions of signal waveforms cannot be expected to match them in detail. The discrepancies between

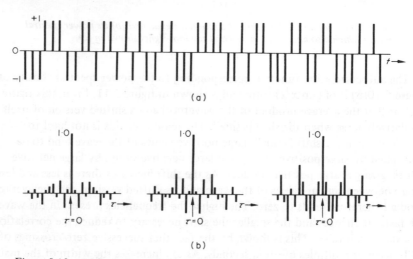

Figure 5.12   *(a) Part of a random sampled-data signal, and (b) three practical estimates of its autocorrelation function*

theoretical and practical estimates of a random signal's properties are often referred to as 'sampling errors'[30]. Errors of this type are illustrated by figure 5.12, which shows a portion of a random sampled-data signal together with typical estimates of its ACF using various numbers of samples. The particular signal chosen has only two possible values, plus or minus 1, with zero mean. Assuming that successive values of the signal are statistically independent, the probability that a given sample will be found to be +1 is just 0·5, and is independent of preceding or following sample values. In this case we would expect the long-term average product of the signal and its shifted version to be zero at all shift values except $\tau = 0$, where it must equal unity. However the estimated ACFs shown in figure 5.12(b), which are based upon signal portions containing 100, 50 and 25 sample values respectively, do not suggest zero correlation at shifts other than zero. The reason for this is that such small portions are not typical of the random waveform as a whole, although the 100-sample portion is much better in this respect than the 25-sample one. Similarly, if the Fourier transforms of the three signal portions were calculated, and their corresponding energy spectra derived by squaring the magnitude of each frequency component. these spectra would be found to display considerable chance fluctuations due to the small number of sample values processed. Indeed it would almost certainly be necessary to smooth the curves obtained before attempting to read significance into them.

With the aid of more advanced theory[30] it is possible to calculate the probability of a given size of sampling error when estimating such properties as the mean, variance, autocorrelation function or power spectrum of a random signal. The difficulty of such a theoretical approach depends very much upon the amplitude probability distribution of the signal, and can be severe. In practice it is often found that hundreds or even thousands of sample values are required if sampling errors are to be reduced to satisfactory levels, and as a general rule the expected size of such errors is approximately inversely proportional to the square root of the duration of the signal portion being processed.

5.5 Important types of random signal

5.5.1 Stationary and ergodic signals

A random signal is described as 'strictly stationary' if none of its statistics is affected by a shift in time origin. A rather wider group of signals are 'weakly stationary', which means that only some of their statistical properties are unaffected by such a shift; these include such measures as the mean, variance, autocorrelation function and power spectrum, which form an adequate description of the signal for many purposes (including the signal processing operations to be described in later chapters). To say that certain statistical properties of a signal are unaltered by a shift in time origin does not mean, of course, that estimates based on limited portions of the signal measured or recorded at different times will be identical—since such estimates are bound to involve sampling errors. On the other hand, in our previous discussion of the statistical properties of random signals, we have assumed that a reliable estimate of some statistical property of a signal may be obtained from a sufficiently long portion of it, regardless of the time origin of that portion. This would not be true of a nonstationary signal. It is worth noting that it may be very difficult to determine whether a signal is stationary or not, particularly in the short term. For the shorter the portion of signal we process, the greater the probability of a large sampling error in the measure we obtain, and the greater the difficulty of knowing whether an apparent change of, say, mean value of a signal from one portion to another reflects a genuine change in signal properties.

The notion of an ergodic process is rather more complicated. In a nutshell, a random signal is described as ergodic if the time-averaged statistics derived from a portion of it are (within the limits imposed by sampling errors) indistinguishable from those derived from a so-called 'ensemble-average'. By ensemble is meant a collection of portions of the signals such as those arranged under one another in figure 5.13. An ensemble statistic is one derived from a set of signal values such as $x_1, x_2, x_3, x_4 \ldots$, one being taken from each member of the ensemble. Therefore if the signal illustrated in figure 5.13 is ergodic, its ensemble statistics would be indistinguishable from those obtained by taking average measures over any one individual signal portion such as A or B. This means that an individual portion must take on all possible values of the signal with the same probabilities as those of the ensemble, and implies that an ergodic process is necessarily a stationary one.

In conclusion, the discussion of random signals presented in this book assumes that useful measures may be obtained by estimating time-averaged properties over a sufficient length of record, regardless of time-origin, and therefore implies stationarity at least in the weak sense described above.

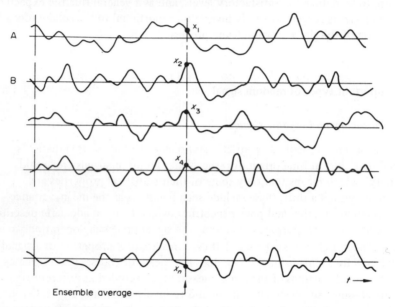

Figure 5.13   *An ensemble of portions of a random signal*

5.5.2 *Gaussian signals*

The gaussian distribution, also known as the 'normal' distribution because of its widespread occurrence in the physical world, was derived by the German mathematician Johann Gauss in the early years of the nineteenth century during his investigations into errors arising in astronomical observations. He considered the total error of a measurement or observation to be the sum of a very large number of very small random errors each of which might be positive or negative in sign. He then showed mathematically that the probability density function of the total error y is of the form

$$p(y) = \frac{1}{\sigma\sqrt{2\pi}} \exp\left(-\frac{y^2}{2\sigma^2}\right)$$

where σ^2 is the variance of the total error variable, σ is its standard deviation, and the constant $(\sigma\sqrt{2\pi})^{-1}$ is included to make the area under the curve equal to unity. In deriving this result Gauss made no attempt to ascertain details of the small contributing errors, and the normal distribution is so general in scope that it has since found a vast number of applications. As far as signal theory is concerned, any random signal which is caused by a number of contributing processes is likely to have a normal amplitude distribution; for example, random noise arising in electronic equipment is very often of this type.

The above probability density curve is symmetrical about $y = 0$, and therefore the mean value of y is zero. In general, however, a normal distribution has a non-zero mean value, denoted by $\bar{y}$, in which case the probability density is given by

$$p(y) = \frac{1}{\sigma\sqrt{(2\pi)}} \exp\left[-\frac{(y - \bar{y})^2}{2\sigma^2}\right]$$

Typical normal probability density functions with different values of mean and variance are shown in figure 5.14. Since the value of a function of the form $\exp(-y^2)$ decreases very rapidly as $|y|$ increases, the probability of finding values of y further than about four standard deviations (4σ) away from the mean value is extremely small. It is indeed often of great interest to know the probability that a signal is greater (or less) than a certain level a, which may be found by integrating the probability density function between the limits a and ∞ (or $-\infty$). With the normal distribution, the integration is not straightforward, but is widely tabulated as the error function (often abbreviated as 'erf'). Values of this function show that, for example, the probability of finding the signal more than 3σ above the mean is 0·00135, and more than 4σ above the mean is 0·00003. Because the normal distribution is of such great practical interest, its properties have been extensively investigated. A very important one is that the sum of two (or more) statistically-independent normal variables is itself normal, with a mean equal to the sum of the means of the contributing processes, and a variance equal to the sum of their variances.

It is sometimes useful to be able to generate a random sampled-data signal with a normal amplitude distribution, for use in, say, a digital computer simulation. As

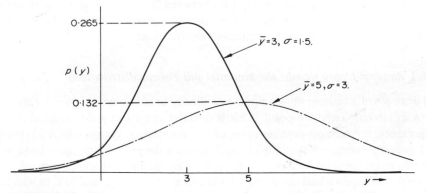

Figure 5.14 *Two gaussian (normal) distributions with different values of mean and variance*

Gauss' result implies, this may be done by adding together a number of independent random variables having any convenient amplitude distribution. To take a very simple example, suppose we conduct a simple statistical test such as throwing a fair die, and record a large number of results. We might get the following sequence

$$5, 6, 1, 1, 3, 5, 4, 2, 5, 1, 6, 5, 3, 4, 6, 2, 4, 3, 2, \ldots \text{etc.},$$

in which the probabilities associated with the six possible values are all equal, and the mean value is 3·5. Next, we form a new sequence of random numbers by adding together groups of, say, twenty consecutive numbers from the above sequence. The minimum possible number in this new sequence will be 20 x 1 = 20, and the maximum 20 x 6 = 120, although both these values would be extremely unlikely. The mean or expected value of the new sequence will be equal to 20 times the mean of the original, or 70, and Gauss' result suggests that the form of its amplitude distribution will be approximately normal. It is also possible to estimate the variance of the new set of numbers as follows. The variance of the original sequence representing the dice throws is

$$\frac{(0\cdot5)^2 + (1\cdot5)^2 + (2\cdot5)^2 + (2\cdot5)^2 + (1\cdot5)^2 + (0\cdot5)^2}{6} = 2\cdot92$$

The variance of the new sequence is then simply equal to 20 x 2·92 = 58·4. The reason for this is that we are summing 20 random sequences to form a new sequence, and it may be shown[15] that variances are additive when the contributing sequences are statistically independent. Hence the standard deviation of the new set is

$$\sigma = \sqrt{(58\cdot4)} = 7\cdot65$$

and it is interesting to note that the maximum and minimum possible values of 20 and 120 lie (120–70)/7·65 = 6·5 standard deviations away from the mean. As we have already seen, such values would lie very much in the 'tails' of the normal curve, and it is therefore reasonable to expect the distribution to be roughly normal in form. However, it should be noted that the distribution is discrete in this case (since it can contain only integer numbers between 20 and 120), and that to realise a true normal probability density curve it would be necessary to sum a large number of random signals with continuous distributions.

5.5.3 Random binary signals: the binomial and Poisson distributions

An example of a random binary signal has already been shown in figure 5.12(a). Such signals have only two possible levels or states and are of great practical importance, for example in those types of communication system which use pulses to transmit information. Although in such systems the probabilities associated with the two levels are often approximately equal, they need not be so; for example, we could envisage a signal consisting of a random series of '1's and '0's in which, on average, the level '1' occurred much less often than the level '0'. Such a signal is illustrated in figure 5.15(a).

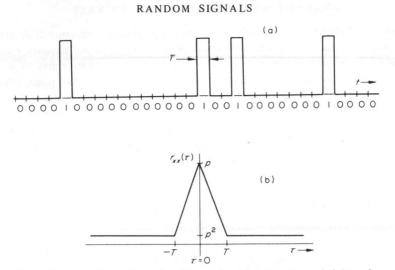

(a)

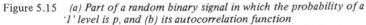

O O O O I O O O O O O O O O O O I O O I O O O O O O O I O O O O

Figure 5.15   (a) Part of a random binary signal in which the probability of a
'1' level is p, and (b) its autocorrelation function

As with other types of random signal, the autocorrelation function (ACF) and power spectral density are important measures which give information about the structure of the function in time and frequency domains. A random binary signal in which there is no systematic relationship between any one signal value and its neighbours has large values of ACF around $\tau = 0$, and a constant value elsewhere which reflects the mean value of the signal. For example, the ACF of the signal of figure 5.15(a) has the form shown in figure 5.15(b), assuming that successive values are statistically independent. The triangular form of the ACF around $\tau = 0$ results from the assumption of square pulses in the time function, and its steady value at all other values of τ represents the chance coincidence of pulses when the signal is multiplied by its shifted version. The corresponding power spectrum would display a zero-frequency term together with a wide distribution of power through the frequency band, the precise nature of which would depend upon the duration (T) of individual pulses in the time function. Basically, such uncorrelated random binary signals are therefore 'wideband'.

The portion of the signal shown in figure 5.15(a) is perhaps typical, but in any other portion of the same length the occurrence of '1' and '0' values would probably be quite different. To take two extreme cases, it is possible that such a portion would contain all '0' values, or possible (although extremely unlikely) that it would contain all '1's'. It might indeed be of considerable practical interest to know the probability that in a sequence of n signal values, there will be r '1's' and $(n - r)$ '0's'. For example, suppose we have designed a radio transmitter to transmit such a signal, the transmitter being turned on whenever a '1' occurs, and off otherwise. Whilst the system would normally have to handle signals in which the '1' level occurred infrequently, it would be sensible to assess the likelihood of its having to handle a sequence of (say) 10 consecutive '1' values, which might well cause it to overheat and fail.

The binomial distribution is relevant to this type of situation, since it describes the probability that, in a sequence of n independent binary signal samples, r take on one of the possible values and $(n - r)$ take on the other, where r may be any integer between 0 and n. In the general case, we will denote the two possible values of the signal as A and B, and denote the probability that any one value is A by the symbol p; then the probability that it is B is just $(1 - p)$. Assuming successive values are statistically independent, we now consider the probability of getting a particular number of 'A's' in a sequence of n values. The probability that each value is A is p^n. The probability that r 'A' values are followed by $(n - r)$ 'B' values is $p^r . (1 - p)^{n-r}$. However, this latter probability applies to every other order of occurrence of r 'A' values in the sequence, and it may be shown[13] that there are $n! \, [r!(n - r)!]^{-1}$ such possible orders. Hence the probability of getting r 'A' values in a sequence of n, regardless of their order, is

$$p_r = \frac{n!}{r!(n - r)!} p^r . (1 - p)^{n-r}$$

which defines the binomial distribution.

Examples of the binomial distribution are given in figure 5.16. When $p = 0.1$ and $n = 10$ (figure 5.16(a)), the probabilities associated with 0, 1, 2 or 3 'A' values are all substantial, although the mean or 'expected' number ($=np$) is the most likely. If $p = 0.5$ the distribution is bound to be symmetrical, as in figure 5.16(b) (this is the situation in a coin-tossing experiment, and the figure therefore represents the probabilities of getting any number of 'heads' or 'tails' between 0 and 10 when a coin is thrown 10 times). Figure 5.16(c) shows the result when $n = 100$ and $p = 0.7$. In this case the individual lines of the distribution are too close together to be drawn individually, so their outline or 'envelope' has been shown instead.

Apart from the application of the binomial distribution to describe the probability that r out of n samples of a two-valued (or 'binary') signal adopt one value rather than the other, there are two limiting cases of particular interest. The first occurs when both n and the expected value (np) are large. Suppose, for example, we have a signal which can take on the numerical values 0 or 1, with $p = 0.5$, and we enquire about the probability that r out of a sequence of 1000

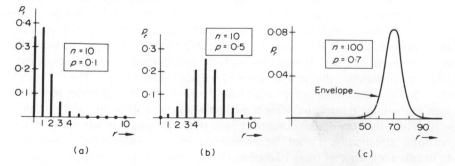

Figure 5.16   *Three examples of the binomial distribution*

values will be equal to 1. The problem is identical to asking the probability distribution of the sum of 1000 values of the signal, and, as we have already seen in the preceding section, we must expect the result to be normally distributed. Thus the normal distribution is one limiting case of the binomial distribution.

The other limiting case of special interest is called the Poisson distribution, which arises when p is very small. Hence it applies to the case of a binary signal in which one of the two possible values occurs very rarely. It may then be shown[13] that the probability of observing r of these rare values in a sequence of length n reduces to

$$p_r = \frac{(np)^r e^{-np}}{r!} = \frac{\mu^r e^{-\mu}}{r!}$$

where μ is the expected value of r. Suppose, for interest, we put $\mu = 1$ and compare this result with the binomial distribution for $p = 0\cdot1$ and $n = 10$ shown in figure 5.16(a). The above expression then becomes

$$p_r = \frac{1^r e^{-1}}{r!} = \frac{1}{r! \, e}$$

which gives the values

$$p_0 = 0\cdot368 = p_1, p_2 = 0\cdot184, p_3 = 0\cdot061, p_4 = 0\cdot015, \text{ etc.}$$

These are fairly close to those shown in figure 5.16(a), and in fact the differences between the binomial and Poisson distributions become negligible for most purposes if $p < 0\cdot01$.

5.5.4 Pseudo-random signals

It is sometimes useful to be able to generate a random signal. Perhaps the most common reason is to allow an accurate model of a physical process in which random phenomena play an important role to be set up on, say, a digital computer. On other occasions, the solution to a problem may only be found by conducting a statistical test which requires a series of random numbers. And finally, we shall see later that it is sometimes convenient to explore the properties of a physical system by applying a random signal to its input and analysing its random response.

Although it is possible to derive a truly random signal from a device such as an electronic noise generator, it is often quite acceptable to use a 'pseudo-random' one. The essential difference between the two is that a pseudo-random signal is generated by using a defined recipe or 'algorithm', and is therefore really deterministic. But since the recipe is a complicated one, examination of a portion of the signal will almost certainly fail to reveal it. In other words the signal looks random, and is effectively so for the experimental situation in which it is to be employed.

As an example of the generation of a pseudo-random signal, suppose we evaluate to nine significant figures the logarithm to base 10 of successive prime numbers 2, 3, 5, 7, 11, 13, 17, etc. The first five values are

$$0 \cdot 301029996$$
$$0 \cdot 477121255$$
$$0 \cdot 698970004$$
$$0 \cdot 845098040$$
$$1 \cdot 041392697$$

We now discard all but the three least significant figures, which we use to represent random numbers between 0 and 999. We expect the numbers to fall anywhere in this range with equal probability. The first twenty pseudo-random numbers of the sequence are plotted in figure 5.17. Using the same set of pseudo-random numbers,

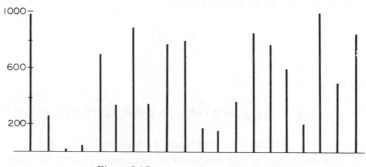

Figure 5.17   *A pseudo-random signal*

a pseudo-random binary signal could easily be generated; for example, a number between zero and some value m is used to indicate one binary level, and any number between m and 999 is used to indicate the other. An appropriate choice of the value for m allows the probabilities associated with the two signal levels to be regulated.

A slightly different form of pseudo-random binary sequence (PRBS) may be generated electronically using what is termed a feedback shift register[21]. Such a device consists, in essence, of a number of interconnected electronic switches with the state ('on' or 'off') of the final switch in the chain indicating the signal level. The switching sequence is normally controlled by a clock which determines the rate of generation of the binary characters. The length of the random sequence is determined by the number (N) of electronic switching stages used, and the way in which they are connected, but is limited to ($2^N - 1$) characters; further clocking causes the sequence to repeat. Although it is possible by using such devices to generate any sequence of binary characters, the so-called 'm-sequences' are of particular interest as pseudo-random signals. In any m-sequence, one of the binary levels occurs once more than the other, so that in a long sequence the probability

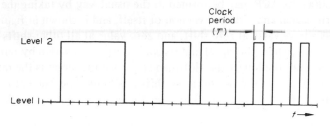

Figure 5.18 *The first 30 binary characters of a 127-character m-sequence*

associated with each level approaches 0·5. This means that an *m*-sequence is
similar to the result of a coin-tossing experiment and indeed it may be shown that
as its length increases, its statistics conform to the binomial distribution. The first
30 characters of an *m*-sequence of length 127 characters ($= 2^7 - 1$) are shown in
figure 5.18.

The general form of the autocorrelation function of a repetitive *m*-sequence is
shown in figure 5.19 for the case where the two signal levels are -1 and $+1$ (with
the former value occurring once more per period than the latter). As in the case of
the signal shown in figure 5.15, the triangular form of the ACF around $\tau = 0$ is due
to the assumption of a signal made up from square pulses of width T, and its
repetition merely reflects the repetitive nature of the sequence. At all other values
of τ, the ACF has a constant value of $-(2^N - 1)^{-1}$, which approaches zero for a
long sequence. As the length of the *m*-sequence is increased, its ACF, and hence its
power spectrum, become more and more like those of the truly random signal
illustrated in figure 5.12.

Of the various other known types of pseudo-random signal, the Huffman
sequence[22] is one of the most interesting from our point of view since it is
designed using z-plane pole–zero concepts. A Huffman sequence is a pseudo-
random sampled-data signal of finite length having particular autocorrelation (and
hence energy spectral) properties, but, unlike the PRBS signals discussed above, it
can take on a wide variety of levels. As an example, consider the sequence shown

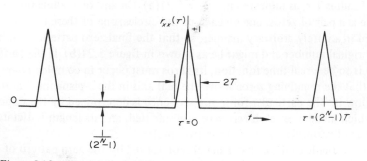

Figure 5.19 *General form of the autocorrelation function of a binary
m-sequence*

in figure 5.20(a). Its ACF may be formed in the usual way by taking the product of the signal and a shifted version of itself, and is shown in figure 5.20(b). The ACF has a large value at zero shift, and zero value at all other shifts except the extreme ones (where the signal and its shifted version overlap by only one sample). In general, the longer the Huffman sequence the larger is the ratio between its values at $\tau = 0$ and at the two extreme shift positions. The value at $\tau = 0$ gives rise to a constant (that is, wideband) term in the energy spectrum, with a relatively

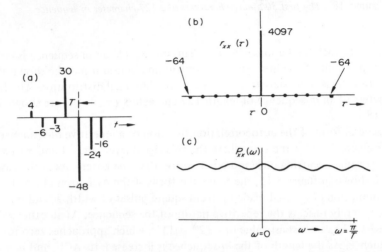

Figure 5.20   *(a) A typical Huffman sequence, (b) its autocorrelation function and (c) its energy spectral density*

small superimposed cosine term due to the two other nonzero ACF terms (see figure 5.20(c)). Since the energy spectra of Huffman sequences are very similar to that of a single Dirac pulse, the sequences are sometimes referred to as 'impulse-equivalent'.

A whole family of sequences having this general form of ACF may be designed by placing two sets of zeros in the z-plane, one on a circle of radius r, the other on a circle of radius $1/r$, as indicated in figure 5.21(a). On any one radius vector such as R there is a pair of zeros, one on each of the circles; one of these now eliminated in a purely arbitrary manner, so that the final zero pattern contains half the original number and might be as shown in figure 5.21(b). If the Huffman sequence is to be a real time function, its zeros must occur in complex conjugate pairs, so that the remaining zeros above the real axis in the z-plane must mirror those below it. The particular form of a sequence is determined by the value of r and by which member of each zero pair is eliminated, and its length is dictated by the number of zeros placed around the two circles.

It is not difficult to show that a signal represented by the zero pattern of figure 5.21(b) has an energy spectrum (or ACF) defined by the zero pattern of figure 5.21(a). The reason for this is that, as we have already seen in section 5.4.3, the power or energy spectrum $P_{xx}(\omega)$ and the spectrum $G(j\omega)$ of a signal are related by

$$P_{xx}(\omega) = G(j\omega)G^*(j\omega) = |G(j\omega)|^2$$

Now any zero on the circle of radius r in the z-plane provides the same contribution to the magnitude of the corresponding spectrum as its 'neighbour' on the circle of radius $1/r$, but an opposite phase contribution (this may be most simply demonstrated by considering their equivalent s-plane locations). Therefore the spectrum $G(j\omega)$ of the Huffman sequence having the zero pattern of figure 5.21(b)

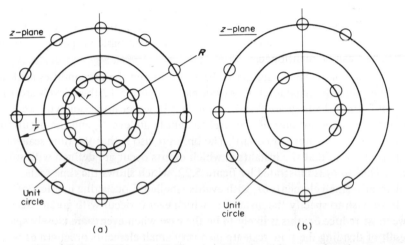

Figure 5.21 *Design of a Huffman sequence by z-plane zero locations*

is the complex conjugate of that defined by the eliminated zero set. Its energy spectrum is essentially 'wideband', since the product of vectors drawn from the various zeros of figure 5.21(a) to a succession of points on the unit circle is itself more or less constant. As the point considered moves around the unit circle, this product will show a slight ripple as successive zero pairs are passed; the ripple, like the one already illustrated in figure 5.20(c), is represented in the ACF by the two finite terms which arise at the two extreme values of the shift parameter τ.

The rather simple example of a Huffman sequence shown in figure 5.20 has 7 sample values and therefore 6 z-plane zeros. In fact it was derived by selecting 6 zeros arbitrarily from two sets placed on circles of radius $\frac{1}{2}$ and 2 in the z-plane.

In this brief account of pseudo-random signals considerable emphasis has been placed upon their power or energy spectra, with the implication that a signal whose energy is fairly evenly distributed over a wide band of frequencies is especially useful. The reason for this is that such signals are, as already mentioned, quite often used as inputs to a physical system under investigation: the relationship between the system's input and output (the latter will also be 'pseudo-random') may only be used to define its behaviour over a wide frequency range if the input itself is 'wideband'. This point will be mentioned again in section 7.4.

5.5.5 *Point process signals*

A point process[23,24] may be defined as one which generates events, and since all such events are assumed to be identical its essential interest resides in their times of occurrence. The word 'event'.embraces a very wide variety of situations; we might define an event as the passage of a car past a certain point on a motorway, or as the generation of an electrical pulse in a nerve fibre, or as the emission of a charged particle from a radioactive source. Clearly, a point process is rather a trivial one if the events occur at strictly regular intervals, and so interesting point process signals are normally random. Fortunately, the work we have already done in previous sections on random and pseudo-random binary signals forms a useful starting point for the brief discussion of point process signals to be presented here. Needless to say, the subject is a large one, and only some of the most basic properties of point process signals can be mentioned.

In our discussion of binary signals we have considered the probabilities associated with the two possible signal levels (for example, in figure 5.12 these levels are $+1$ and -1, and in figure 5.15 they are 0 and 1). There is however no reason why we should not use the occurrence of one level to denote an 'event', and the occurrence of the other level to denote 'no event'. The only proviso is that the time scale be divided up finely so that the instants at which events occur are defined with sufficient accuracy. This point is illustrated by figure 5.22, which shows the time scale divided up into small elements δt, with events labelled E occurring in a few of them. If we wish to specify the instants at which events occur with greater accuracy, then we must reduce δt; this is likely to be the case when events are closely spaced. The result of dividing the time scale up into very small elements of length δt is that the probability of an event occurring in any one such element becomes very small. This means that in cases where the time of occurrence of any one event is completely unaffected by those of its neighbours the process is described by the Poisson distribution.

Examples of the Poisson distribution as a description of a random point process are fairly common. Suppose, for example, we stand on a bridge over a motorway and observe vehicles passing underneath, which we class as 'events'. Provided that the motorway is not busy, we may expect events so defined to be independent of one another; in other words, vehicles are not restrained by those in front or behind, and do not therefore bunch together. In these conditions, we could only sensibly allocate a small and constant value to the probability that a vehicle passes under

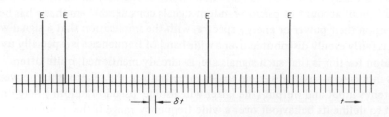

Figure 5.22   *A random event sequence*

the bridge during any small time element δt. Hence the probability of observing r events in a time interval of length T_0 is expected to conform closely to the Poisson distribution

$$p_r = \frac{\mu^r e^{-\mu}}{r!}$$

But μ, the expected number of events, is just equal to pn, where n is the number of elements of length δt within T_0, and p is the probability of an event occurring within any one such element. Hence

$$\mu = pn = p \cdot \frac{T_0}{\delta t}$$

Furthermore if we define λ as the average rate of the process (measured in events per second), then

$$\lambda = \frac{pn}{T_0} = \frac{\mu}{T_0} = \frac{p}{\delta t}, \qquad \text{and} \qquad \mu = \lambda T_0$$

The Poisson distribution might be described as the most parsimonious of all possible point-process distributions, in the sense that it is completely defined by its mean rate, λ. No other distribution may be defined by just one parameter. Therefore if we have no information about a point process other than its mean rate, and if there is no evidence of constraints between successive events, however defined, then we may expect it to be a Poisson process.

Another matter of great interest when discussing a random point process is its interval distribution, which describes the probability that a time interval between two successive events is of a certain length. We first evaluate the interval distribution of a Poisson process. Suppose an event has just been observed; the probability that the next event will be observed during the following δt is just p (and is unaffected by the recent occurrence of the first one); this is therefore also the probability that the interval is of length δt. If the interval is to be of length $2\delta t$, two conditions must be met; there must have been no event in the first element δt, and there must be an event in the second. The associated probability is therefore $(1 - p) \cdot p$. A simple extension of this argument shows that the probability associated with an interval of $n \cdot \delta t$ is

$$p_n = (1 - p)^n p$$

Successive values of p_n are plotted in figure 5.23, for the case when $p = 0.03$; they lie on a decaying exponential curve. As δt is reduced, the form of this curve remains unaltered, although its vertical scale changes. In the limit it becomes a probability density function.

As we have seen, the occurrence of an event in a Poisson process gives no clue whatsoever to when the next event is likely to occur; on other occasions, however, we are faced with a process which is rather more regular, such as the beating of a heart. In such a case, there is clearly a relatively small preferred range of intervals,

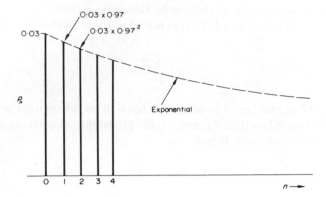

Figure 5.23 *Probabilities associated with intervals of length (n . δt) in a Poisson process having a mean rate of (0·03/δt) events per second*

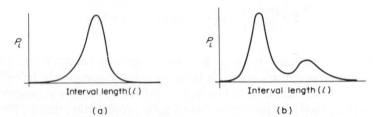

Figure 5.24 *Interval distributions of point processes with (a) one and (b) two preferred ranges of intervals*

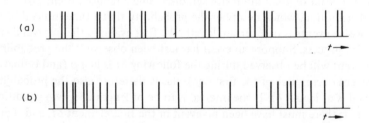

Figure 5.25 *Portions of two point processes having identical interval distributions*

and the interval distribution may typically resemble that of figure 5.24(a). Knowledge of the instant of occurrence of the last heartbeat would clearly allow a sensible prediction about when the next one should be expected. Figure 5.24(b) illustrates a process having two preferred interval ranges; this could arise, for example, if the occasional event in a fairly regular train of events was missing or deleted. It is clear that neither of these interval distributions (unlike that of figure 5.23) could be defined merely by its mean value.

The interval distribution of a point process gives no information about any possible correlation between successive intervals and is therefore known as a 'first-order' measure. For example, figure 5.25 shows portions of two point processes; one is a Poisson process of the type already discussed, in which there is no systematic relationship between the size of any one interval and neighbouring ones; in the other, intervals either shorter or longer than the average tend to occur in groups. The two signals have identical interval distributions, and in order to quantify the relationships between successive intervals it would be necessary to use extra statistical measures. For example a serial correlation coefficient ρ_k may be defined as

$$\rho_k = \lim_{N \to \infty} \frac{1}{(2N + 1)} \sum_{m=-N}^{N} \frac{(T_m - \bar{T})(T_{m+k} - \bar{T})}{(T_m - \bar{T})^2}$$

where $\bar{T}$ is the average interval and T_m represents the mth interval in a sequence. ρ_k is therefore a long-term average measure of the product of departures from the mean of the mth and $(m + k)$th interval, divided by the variance of the interval sequence, and like other correlation functions is a 'second-order' measure. If there is no systematic relationship between the mth and $(m + k)$th intervals, ρ_k is zero; if they both tend to be either longer or shorter than the mean, ρ_k is positive; and if one is shorter and the other longer than the mean, ρ_k is negative. The manner of estimating successive values of ρ_k is similar to that for the set of correlation coefficients which describe the relationship between a sampled-data signal and its shifted version (see section 5.4.2); however, in this case we are talking about correlation between the sizes of neighbouring intervals, rather than neighbouring signal levels, and ρ_k is not therefore a linear function of time-shift.

A more conventional type of correlation function may be estimated for a point-process signal, by treating events as narrow (Dirac) pulses arranged along the time axis. If we autocorrelate a limited portion of such a time function, the result will itself be a function composed of Dirac pulses as shown in figure 5.26, with every pulse in the ACF denoting the coincidence of a pulse in the original sequence with one in its shifted version. In the limit of a very long signal record, the pulses in the ACF become very numerous and it is sensible to talk about the density of pulses

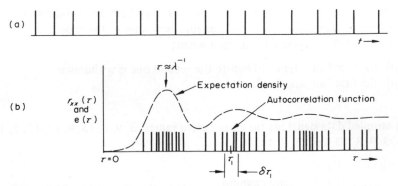

Figure 5.26 *(a) A portion of a fairly regular random point process, and (b) its autocorrelation function for positive values of time shift; superimposed is the expectation density function, e(τ)*

in any small shift increment such as $\delta\tau_1$ centred on τ_1. For convenience, the large value of ACF at $\tau = 0$ is often omitted, and the resulting function called the 'expectation density function', $e(\tau)$. Essentially a further example of a probability density function, the expectation density describes the probability of an event occurring τ seconds after some reference event, as a function of τ. If events are spaced more or less regularly in time, $e(\tau)$ displays a number of humps corresponding to the first, second, and further events which follow the reference event. As τ becomes large, the expectation density curve flattens out to a level which reflects the mean rate of the process. The relevant expectation density curve is superimposed upon the ACF in figure 5.26.

In this brief outline of point-process signals, an attempt has been made to indicate some of the more important measures relevant to random event sequences. Although first-order measures such as the interval distribution are important, in practice it is nearly always second (or higher) order statistics such as correlation functions which prove of most interest. Not surprisingly, most of these measures are interrelated rather as are the time and frequency domain descriptions of more conventional signals, but the precise nature of the relationships are rather complicated and may only be revealed by a more comprehensive theoretical approach[23].

Problems

1. Two fair dice are thrown; assign probabilities to the events defined by:

   (i) a '3' on one die accompanied by a '4' on the other.
   (ii) a number less than 3 on one die accompanied by a number greater than 3 on the other.
   (iii) the sum of the two dice scores being either 11 or 12.

2. A box contains 3 red balls and 2 green ones. Two balls are withdrawn in succession. Assess the number of possible outcomes of the experiment and the probability associated with each when:

   (i) the first ball is replaced before the second one is withdrawn.
   (ii) the first ball is not replaced.

3. A sampled data signal has the consecutive values: 3, 9, 6, 15, 4, 13, 12, 1, 11, 4, 3, 9, 8, 12, 2, 7, 1, 7, 14, 10.

   Estimate: (i) its mean
   (ii) its second moment
   (iii) its second central moment (that is, variance), and standard deviation.

4. The accompanying figure shows a triangular probability density function relating to the amplitude levels of a random signal.

   (i) find b in terms of a so that the function is correctly normalised.
  (ii) estimate the mean and standard deviation of the signal.
 (iii) sketch the cumulative distribution function.
 (iv) estimate the probability that x is greater than $a/2$.

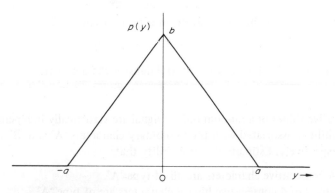

5. A continuous random signal is always positive, and is described by the amplitude probability density function:

$$p(y) = y\, e^{-y}$$

in the range $0 < y < \infty$

Calculate   (i) its mean and variance
           (ii) the probability that $y > 1$
          (iii) the probability that $1 < y < 2$

6. An isolated rectangular pulse of duration T seconds and unit amplitude is centred on $t = 0$. Derive its autocorrelation function and energy spectrum. Show that the latter is simply the square of the pulse's frequency spectrum, and explain.

7. Comment on the usefulness of the autocorrelation function (ACF) for the description of a random signal.
    The ACF of a signal waveform containing both random and periodic components is given by

$$r_{xx}(\tau) = A\, e^{-b|\tau|} + \cos \omega_0 \tau,$$

where A, b, and ω_0 are constants.

Sketch a typical portion of the waveform and estimate:

(i) the mean square value of the signal
(ii) the standard deviation of its random component.
(iii) the frequency at which the power spectral density of its random component is reduced to $(1/\sqrt{2})$ of its value at $\omega = 0$.
(iv) the frequency of its periodic component.

8. The ACF of a random noise waveform is found to be

$$r_{xx}(\tau) = A\ e^{-b|\tau|}\ \cos \beta\tau$$

Calculate and sketch its power spectral density characteristic.

9. Successive values of a random binary signal are statistically independent. The probabilities associated with the two binary characters 'A' and 'B' are 0·2 and 0·8 respectively. Estimate the probability that

(i) 4 consecutive characters are all of type 'A'.
(ii) any 2 of 4 consecutive binary characters are of type 'A'.
(iii) the sequence 'AAA' occurs somewhere in a portion of signal of length 10 characters.

10. Derive the Huffman sequence corresponding to the z-plane zero plot in the adjacent diagram. Find its ACF for all relevant values of shift, and sketch its energy spectrum.

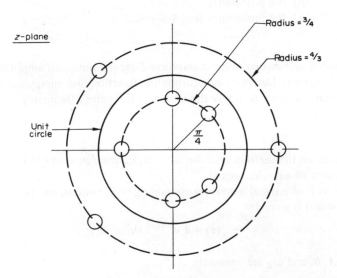

11. The arrival of customers at a shop may be described as a stationary random point process with Poisson statistics. If the average arrival rate is 100 customers per hour, estimate the probability that in any one-minute period

 (i) no customers arrive,

(ii) more than 3 customers arrive.

If it takes 20 seconds to serve each customer, what is the probability that any customer is kept waiting by one or more persons in front of him?

6

Signal Comparison

6.1 Introduction

Although we have so far concentrated on the problem of describing or analysing individual signals, the foregoing chapters have in fact laid much of the groundwork for a discussion of the ways in which signals may be compared with one another.

Any of the measures of signal properties so far described may be used as a basis for comparison. For example, if we have two random signals, we may compare their amplitude distributions, mean values, variances, autocorrelation functions, and so on. Although it is sometimes adequate to make a subjective assessment of the similarities of two signals, such as that they have 'approximately equal mean values', it is often useful to have a definite quantitative measure of shared properties. Fortunately, the description of orthogonal functions and their applications given in section 2.3 forms a good introduction to the derivation of useful quantitative measures.

In section 2.3.1, where the analogy between vectors and signals was discussed, we saw how a signal $f_1(t)$ could be approximated over any interval $t_1 < t < t_2$ by another signal or waveform $f_2(t)$, together with an error term $f_e(t)$

$$f_1(t) = C_{12} . f_2(t) + f_e(t)$$

We then adjusted the coefficient C_{12} to achieve some desired error criterion— normally to minimise the average squared value of $f_e(t)$ in the interval $t_1 < t < t_2$ (which is the same as minimising the 'error power'). This particular criterion allowed the appropriate value of C_{12} to be expressed quite simply as

$$C_{12} = \frac{\int_{t_1}^{t_2} f_1(t) . f_2(t) . dt}{\int_{t_1}^{t_2} f_2^{2}(t) . dt}$$

112

Thus if the product of $f_1(t)$ and $f_2(t)$ in the stipulated interval is zero, we say that there is no component of $f_2(t)$ in $f_1(t)$; in other words the signals are orthogonal. Conversely, a large value of C_{12} denotes the fact that $f_1(t)$ and $f_2(t)$ have much in common. We have therefore defined a quantitative approach towards comparison of the two signa... $f_1(t)$ and $f_2(t)$.

Although this general concept could be used to compare various aspects of two signals—for example, their amplitude probability density functions—by far the most common practice is to use it, as in the above discussion, to compare the time functions describing their actual waveforms. As a starting point, let us assume that

$$f_1(t) = A_1 \cos \omega_1 t \qquad \text{and} \qquad f_2(t) = A_2 \cos \omega_2 t$$

and let us use as a measure of their relationship the average product over a long interval T_0, given by

$$\lim_{T_0 \to \infty} \frac{1}{T_0} \int_{-T_0/2}^{T_0/2} f_1(t) \cdot f_2(t) \cdot dt$$

(this measure is no different in principle from C_{12} defined above; we have merely chosen not to normalise by the mean square value of $f_2(t)$, and to take an averaged version of the result over a long time interval). Because of the orthogonality of cosines of different frequency, our measure will be zero unless $\omega_1 = \omega_2$. If we now change $f_2(t)$ to $A_2 \sin \omega_2 t$, the result will be zero for every value of ω_2 including $\omega_2 = \omega_1$, because of the orthogonality of sine and cosine waveforms. Hence our comparative measure of $f_1(t)$ and $f_2(t)$ will only be nonzero by virtue of shared cosine, or sine, components at particular frequencies; unfortunately, however, as it stands it will yield no information about which frequencies are common to the two waveforms or about their relative amplitudes. This limitation may be resolved by use of the rather more comprehensive measure known as the cross-correlation function.

6.2 The cross-correlation function

The cross-correlation function relating two signals $f_1(t)$ and $f_2(t)$ may be defined as

$$r_{xy}(\tau) = \lim_{T_0 \to \infty} \frac{1}{T_0} \int_{-T_0/2}^{T_0/2} f_1(t) \cdot f_2(t + \tau) \cdot dt$$

where τ is a time-shift imposed upon one of the signals. Two important points may be noted at once: firstly, the cross-correlation function (CCF) and the auto-correlation function (ACF) described in the last chapter are very similar, the only difference being that the CCF applies to two different signals, whereas the ACF relates a signal to a shifted version of itself: and secondly, the definition of the CCF ties in closely with the above discussion on orthogonality and signal comparison. However, instead of deriving a single index (such as the coefficient

C_{12}) to describe the similarity between two signals, we now define a variable r_{xy} which is a continuous function of the imposed time-shift τ. When the signals to be compared are of limited duration it is appropriate to use a 'finite' version of the cross-correlation function, given by

$$r_{xy}(\tau) = \int_{-\infty}^{\infty} f_1(t) . f_2(t + \tau) . dt$$

This exactly parallels the use of the 'finite' version of the ACF mentioned in section 5.4.2.

The benefits of introducing a time shift between $f_1(t)$ and $f_2(t)$ are not hard to appreciate. Suppose, to take a rather extreme example, $f_1(t)$ and $f_2(t)$ are two random signals which differ only in their time origin, as illustrated in figure 6.1(a). The average of the product $[f_1(t) . f_2(t)]$ over a long time interval may well

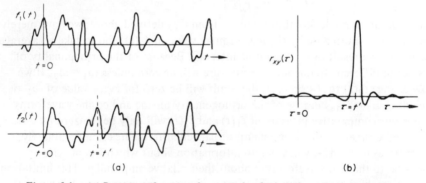

Figure 6.1   (a) Portions of two random signals which differ only in their time origin, and (b) an estimate of their cross-correlation function

be zero, but we would be quite wrong to infer from this that they have nothing in common; indeed, apart from a simple time shift of t' seconds, they are identical. This fact is revealed clearly by their CCF, the value of which is large when the shift τ equals t', as shown in figure 6.1(b). Similarly, a sine and cosine wave of the same frequency, which are orthogonal and therefore 'have nothing in common' as they stand, certainly do have something in common when one of them is shifted. In other words the cross-correlation function satisfies the commonsense notion that two signal waveforms with common frequency components are in an important sense alike, in spite of different timing or phase relationships.

In order to formalise the properties of the cross-correlation function we now investigate its form for the two signals

$$f_1(t) = A_1 \cos \omega t, \qquad f_2(t) = A_2 \cos(\omega t + \theta)$$

where θ represents a general value of phase shift.

We have

$$r_{xy}(\tau) = \lim_{T_0 \to \infty} \frac{1}{T_0} \int_{-T_0/2}^{T_0/2} A_1 A_2 \cos \omega t . \cos [\omega(t + \tau) + \theta] . dt$$

$$= \lim_{T_0 \to \infty} \frac{1}{T_0} \int_{-T_0/2}^{T_0/2} \frac{A_1 A_2}{2} [\cos(2\omega t + \omega \tau + \theta) + \cos(\omega \tau + \theta)] . dt$$

The first cosine term in the integrand averages out to zero over a long interval; the second is not a function of t, hence

$$r_{xy}(\tau) = \lim_{T_0 \to \infty} \frac{1}{T_0} . \frac{A_1 A_2}{2} . \cos(\omega \tau + \theta) . \left[t \right]_{-T_0/2}^{T_0/2}$$

$$= \frac{A_1 A_2}{2} \cos(\omega \tau + \theta)$$

Thus the form of the CCF reflects the product of the amplitudes of $f_1(t)$ and $f_2(t)$, their common frequency ω, and their relative phase angle θ; it is plotted in figure 6.2. When the two signals being cross-correlated share a number of common frequencies, each gives a corresponding contribution to the CCF. In this respect the auto- and cross-correlation functions are essentially similar; however, the ACF loses all information about the phase of various frequency components in a signal, whereas the CCF retains information about the relative phases of common frequency components in the two signals which it is comparing. A consequence is that, unlike the ACF, the CCF is not normally an even function of τ.

Since the cross-correlation function is essentially a time-averaged measure of shared signal properties it is very suitable for comparing random signals. It then yields an average measure of frequency components held in common, and a practical estimate of it only approaches the theoretical one when very long portions of the signals are available; otherwise there are the inevitable 'sampling errors'

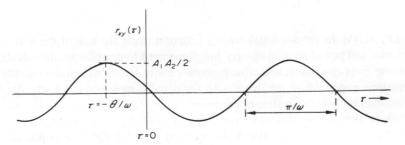

Figure 6.2 *The cross-correlation function for two cosine waves of amplitudes A_1 and A_2, frequency ω, and relative phase angle θ*

associated with any attempt to extract statistical measures from limited signal portions (see section 5.4.4). An example of a cross-correlation function applied to random signals has already been shown in figure 6.1, illustrating the great use of the CCF in showing up time delays. A further example is shown in figure 6.3, which represents an estimated CCF of two random signals with shared frequency ranges. The periodicities in the CCF reflect these shared ranges, although it is certainly difficult to extract any precise information from the figure. In cases such as this where simple time shifts are not involved, it is often more appropriate and rewarding to use the so-called cross-spectral density function.

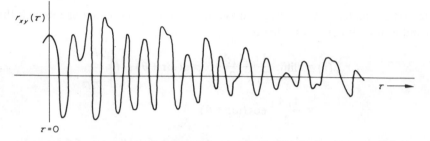

Figure 6.3   *A typical estimated cross-correlation function for two random signals with two closely spaced, shared, frequency ranges*

6.3   The cross-spectral density function

Just as the autocorrelation function of a signal has its counterpart in the frequency domain, so the cross-correlation function may be transformed into an equivalent frequency domain function—the cross-spectral density, or cross-spectrum. Formally, the relationship is expressed in terms of the following Fourier transform pair

$$P_{xy}(\omega) = \int_{-\infty}^{\infty} r_{xy}(\tau) . e^{-j\omega\tau} . d\tau$$

and

$$r_{xy}(\tau) = \frac{1}{2\pi} \int_{-\infty}^{\infty} P_{xy}(\omega) . e^{j\omega\tau} . d\omega$$

where $P_{xy}(\omega)$ is the cross-spectral density function. As in the case of the auto-correlation and power spectral density functions used to describe an individual signal, the cross-correlation and cross-spectral density functions contain equivalent information, and which of the two is used for signal comparison in a particular case is therefore largely a matter of convenience.

We have already seen that when two signals $f_1(t) = A_1 \cos \omega t$ and $f_2(t) = A_2 \cos(\omega t + \theta)$ are cross-correlated the resulting wave has the same period in τ as $f_1(t)$ and $f_2(t)$ have in t, with an amplitude equal to $A_1 A_2/2$. Furthermore, its maximum value occurs at $\tau = -\theta/\omega$, where θ is the relative phase shift between the two underlying signals (see figure 6.2). The corresponding spectral description therefore involves a term of magnitude $(A_1 A_2/2)$ and phase angle θ, at a frequency ω radians/second. In the general case, $f_1(t)$ and $f_2(t)$ contain a number of common frequencies or frequency ranges, all of which are indicated by their cross-spectrum. A typical cross-spectrum of two periodic signals containing a number of common components is shown in figure 6.4: each cross-spectral term has a magnitude equal

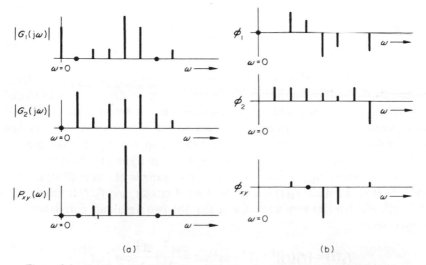

(a)   (b)

Figure 6.4   *(a) The magnitude characteristics of two signal spectra and of their cross-spectrum, and (b) the corresponding phase characteristics*

to half the product of the amplitudes of the corresponding waves in $f_1(t)$ and $f_2(t)$, and a phase equal to the difference between their individual phase angles.

Unlike the power spectrum of a single signal, the cross-spectrum relating two different signals is generally complex, involving phase as well as magnitude terms as illustrated above in figure 6.4. In the rather special case when $f_2(t)$ is essentially a delayed but otherwise unaltered version of $f_1(t)$, as in figure 6.1, the phase difference between common components is just proportional to their frequency (this point has already been covered in section 4.2), whereas their amplitudes are the same. Therefore the cross-spectrum will have a magnitude characteristic equal to that of the power spectrum of each individual signal, and a phase characteristic which is proportional to frequency.

It is sometimes convenient to split the cross-spectrum function $P_{xy}(\omega)$ into real and imaginary parts, rather than work with magnitude and phase characteristics. Hence

$$P_{xy}(\omega) = C_{xy}(\omega) + j\, Q_{xy}(\omega)$$

where $C_{xy}(\omega)$ is called the 'co-spectrum' and $Q_{xy}(\omega)$ the 'quadrature-spectrum'. We therefore have

$$|P_{xy}(\omega)| = [C_{xy}^2(\omega) + Q_{xy}^2(\omega)]^{1/2}$$

and

$$\phi_{xy}(\omega) = \tan^{-1}\left[\frac{Q_{xy}(\omega)}{C_{xy}(\omega)}\right]$$

where $\phi_{xy}(\omega)$ represents the phase difference between components in the two signals at a frequency ω.

6.4   Discussion

In view of the central importance of frequency spectra in signal theory, it is not surprising that the two indices most commonly used for signal comparison—the cross-correlation and cross-spectrum functions—are essentially measures of shared spectral components. Their widespread use should not, however, obscure the fact that it is occasionally useful to compare other aspects of two signals, such as their amplitude distributions or moments. It must also be appreciated that the cross-correlation function fails to indicate any correlation in some situations where commonsense would suggest otherwise. A simple example of this is illustrated in figure 6.5. Here the signal $f_1(t)$ is a cosine wave $A \cos \omega t$, and $f_2(t)$ is simply equal to $[f_1(t)]^2$, which might very well suggest that the two signals are correlated. However, since

$$f_2(t) = [f_1(t)]^2 = A^2 \cos^2 \omega t = \frac{A^2}{2} + \frac{A^2}{2} \cos 2\omega t$$

$f_2(t)$ contains both a zero-frequency component and one at 2ω radians/second, but not one at ω radians/second. Therefore $f_1(t)$ and $f_2(t)$ have no shared frequencies and their cross-correlation function is zero for all values of τ. The basic reason for this is that the relationship between $f_1(t)$ and $f_2(t)$ is a so-called 'nonlinear' one, whereas the correlation functions we have described in both this and the previous chapter indicate only 'linear' signal relationships. The meaning

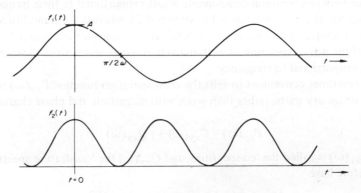

Figure 6.5

of the terms linear and nonlinear in a signal context will be explored rather more fully in the next chapter.

The practical importance of the cross-correlation and cross-spectrum functions is easier to appreciate if we consider why we might wish to compare two signals in the first place. The most likely reason is that some sort of causal relationship between them is anticipated, and we wish to define it quantitatively. For example, the signal $f_1(t)$ might represent an input or disturbing function applied to some system (such as a mechanical force exerted on a structure), and $f_2(t)$ a response (for example, a resulting mechanical displacement) measured at some point. If we are able to define the relationship between $f_1(t)$ and $f_2(t)$ by cross-correlating them, it is not hard to see that we have obtained an important measure of the properties of the system itself. Furthermore, since the cross-correlation function applies to random as well as to deterministic signals, it suggests the possibility of defining the performance of a system by disturbing it with a random input. However at this stage we are beginning to trespass heavily on the subject of signal processing, rather than signal analysis, and a fuller discussion of these points is therefore best postponed until some basic aspects of signal processing have been introduced.

Problems

1. Two random signals have equal mean values and variances. What clues, if any, does this information give about similarities between their

   (i)   average powers,
   (ii) autocorrelation functions,
   (iii) second moments,
   (iv) amplitude distributions,

   and what clues does it give about their

   (v)   cross-correlation function,
   (vi) cross-spectrum?

2. Evaluate the cross-correlation function (CCF) of the square and sinusoidal waveforms shown in the accompanying figure, and show that it reflects just

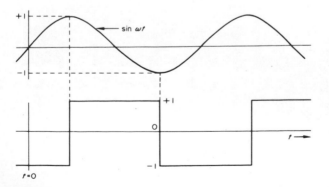

their single common frequency component. Comment on the peak magnitude and phase of the CCF.

3. A random signal is known to have the autocorrelation function shown in part (a) of the accompanying figure: a second random signal, statistically independent of the first, has the power spectrum indicated in part (b). What is the theoretical form of their cross-spectrum? What may be inferred about their cross-correlation function? Would these two measures be of any practical interest?

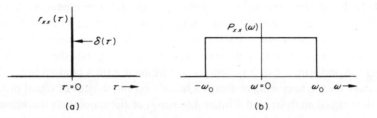

(a)   (b)

PART II

SIGNAL PROCESSING

7

Signals and Systems

7.1 Introduction

Earlier chapters of this book have concentrated on the ways in which signals may
be described and analysed, and, as a natural extension of the discussion, we now
turn our attention to the ways in which signal properties are modified by various
types of processing. Some of the major reasons for interest in signal processing
were outlined in sections 1.3 and 1.4. There it was mentioned that by far the
majority of signal processing devices are electrical or electronic, typical examples
being the radio or radar receiver, telephone-exchange equipment, and the electronic
computer. Fortunately it is not necessary to have any detailed understanding of
electrical circuits or elements to appreciate the main concepts of signal processing.
The approach adopted in this and following chapters is to consider a signal
processing device simply as one which delivers to its output a modified version of
a signal applied to its input. In other words, it is regarded as an 'input–output'
device, whose detailed internal construction does not concern us. This is often
referred to as the 'black-box' approach: it involves us in mathematical descriptions
of the overall performance of the signal processor, but not of the detailed electrical
circuits of which it is probably constructed.

The scope of signal processing and the types of question which we shall seek to
answer in the following chapters may be illustrated by a few examples. We might
wish to predict the effect of a signal processing device or operation on either the
waveform or on the frequency spectrum of a particular signal: or to define the
characteristics of a processing device required to achieve a desired modification to
the signal. Another important type of problem arises when a signal is mixed with
an unwanted disturbance, often of a random nature, and the signal is to be
processed so as to reduce that disturbance as much as possible. And finally, it is
often valuable to observe a signal at two or more points in a processing system and,
by comparing them, to deduce something about the system's characteristics.

7.2　Basic aspects of linear processing

7.2.1　Linear systems

There is an extremely important class of systems, which includes many signal processing devices of great interest, known as 'discrete linear' systems. The theory of such systems has been extensively developed and forms a coherent and satisfying body of knowledge, with important applications in such fields as electric circuits, signal theory, automatic control and mechanical vibrations.

In practice, such systems are formed by the interconnection of discrete linear elements: amongst such elements are included the familiar resistors, inductors, and capacitors used in electric circuits, and the mass, spring, and viscous damper elements commonly found in mechanical systems. Electronic devices such as valves and transistors may also be regarded as linear in many of their most common signal processing applications. The reader who has some familiarity with differential equations will know that such systems may be described by so-called linear differential equations with constant coefficients.[8] Typically, such an equation takes the form

$$c_n \cdot \frac{\mathrm{d}^n f_2(t)}{\mathrm{d}t^n} + c_{n-1} \cdot \frac{\mathrm{d}^{n-1} f_2(t)}{\mathrm{d}t^{n-1}} + \ldots c_1 \cdot \frac{\mathrm{d}f_2(t)}{\mathrm{d}t} + c_0 \cdot f_2(t) = f_1(t)$$

although in the most general case the right-hand side of the equation includes derivatives of $f_1(t)$ as well. $f_1(t)$ is an input or 'forcing function' applied to the system, and $f_2(t)$ is some response variable of interest, such as a voltage at some point in an electrical system or a velocity at some point in a mechanical one. The coefficients $c_n, c_{n-1}, \ldots c_0$ are constants related to the values of the discrete linear elements of which the system is composed. In the signal processing context, $f_1(t)$ and $f_2(t)$ may be thought of as the input and output signal waveforms of a processing device. Any circuit, system, or device which may be described by a differential equation of this general form is called 'linear'. Great care must however be exercised in the use of this word, because it is certainly not meant to imply a simple straight-line relationship between any two variables of the system; for example, the instantaneous values of the input and output signals of a linear system are not in most cases linearly related.

The term 'linearity' when applied to systems or signal processing devices does however denote other major properties, amongst which that of superposition is probably the most important. The principle of superposition states that the output or response of a linear system due to several inputs simultaneously applied is equal to the sum of its responses to each input applied separately. Thus if an input $f_1(t)$ gives rise to an output (or measured variable at any point) denoted by $f_2(t)$, and if input $f_1'(t)$ produces output $f_2'(t)$, then the combined input $[f_1(t) + f_1'(t)]$ will give rise to the output $[f_2(t) + f_2'(t)]$. A corollary is that an input equal to $A \cdot f_1(t)$ gives rise to an output $A \cdot f_2(t)$, where A is a constant. Although at first the principle of superposition may seem somewhat trivial, its implications for signal processing are in fact profound. For example, when superposition applies we may consider any signal input to be made up from a number of separate components—such as sine and cosine waves—and evaluate the response to each component; the output signal is then found by summation of the individual responses. This possibility gives enormous power to concepts such as Fourier analysis described in earlier chapters of this book.

A second major property of linear systems is what may be termed 'frequency-preservation'. This means that the output, or response, of the system to any input contains only those frequencies or frequency ranges represented in the input itself. Thus no new frequencies are generated within the system; all that such a system does is to modify the amplitudes and relative phases of the various input components before delivering them to the output. We have already seen that Fourier analysis techniques (and their extension in the form of the Laplace transform) allow us to represent any practical signal waveform as a sum of a number of frequency components. If the signal is now applied as an input to a linear signal processing device, its various components will be modified in amplitude and phase before reaching the output; the latter is, by the principle of superposition, equal to the sum of all such modified components. Thus if we can define the way in which different frequencies are modified in passing through the system, we have a powerful method of defining its response to any signal waveform.

The reason for this important frequency-preservation property is the form of the differential equation relating input and response of a linear system. It may be shown[8] that if the input or forcing function $f_1(t)$ is of the form e^{st}, then every measurable time-varying quantity (including the response of interest) in the system is also of exponential form, with the same index s. Terms of the form e^{st} include real exponentials, imaginary exponentials (which, in pairs, constitute sine and cosine waves) and complex exponentials (which, in pairs, represent decaying or increasing sine or cosine waveforms): therefore any of these waveform types is preserved in a linear system. It is worth noting that the Laplace transform, introduced in section 3.4, effectively analyses a signal waveform into components of the form e^{st}, and indeed a number of waves of this general type have already been illustrated in figure 3.11. We must therefore expect that the Laplace transform has a special relevance for the discussion of signal processing by linear systems.

7.2.2 The frequency-domain approach

Suppose a continuous sinusoidal signal $A \sin(\omega_0 t - \theta)$ forms the input to a linear processing device. The output will be a wave at the same frequency, but with modified amplitude and phase, typically as shown in figure 7.1. In the general case, the input signal may be considered to be made up from a large number of frequency components, and the modifications to amplitude and phase caused by the system will be frequency-dependent. Let the spectrum of the input signal be denoted by $G_1(j\omega)$ (which defines both its magnitude and phase characteristics) and let the modifications to this spectrum caused by the system be denoted by the complex function $H(j\omega)$. At any value of ω, $H(j\omega)$ also has both magnitude and phase terms; the former represents a multiplication factor and the latter an imposed

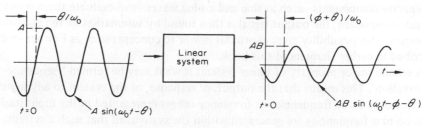

Figure 7.1

shift. For example, the linear system of figure 7.1 magnifies a wave of frequency ω_0 by a factor B and imposes a phase lag of ϕ radians; hence

$$H(j\omega)\,|_{\omega=\omega_0} = H(j\omega_0) = B\,e^{-j\phi}$$

$$= B\cos\phi - jB\sin\phi$$

and

$$|H(j\omega_0)| = \sqrt{(B^2\cos^2\phi + B^2\sin^2\phi)} = B$$

Furthermore, if we denote the spectrum of the output signal by $G_2(j\omega)$, then

$$G_2(j\omega) = G_1(j\omega)\,.\,H(j\omega)$$

This important expression relies upon the rule for multiplication of complex numbers: if we have two complex numbers denoted by $r_1\exp(j\theta_1)$ and $r_2\exp(j\theta_2)$, then their product is simply $r_1 r_2\exp[j(\theta_1 + \theta_2)]$. The magnitude $(r_1 r_1)$ of their product is therefore the product of their individual magnitudes, associated with a phase angle $(\theta_1 + \theta_2)$ which equals the algebraic sum of their individual phases. The output spectrum $G_2(j\omega)$ is hence found by multiplying the magnitude characteristics of $G_1(j\omega)$ and $H(j\omega)$, and adding their phase characteristics. It is worth noting that the same output signal spectrum would be obtained if an input signal with spectrum $H(j\omega)$ was applied to a linear system described by the function $G_1(j\omega)$. Hence a complex frequency function may describe either a signal or a linear system; in this sense signals and systems are analogous.

The complex function which defines the operation of a linear system as a function of frequency is called its 'frequency response'. The great attraction of the frequency-domain approach is that, as we have just shown, the spectrum of the output signal or waveform is equal to the product of that of the input and the system's frequency response. It is important to realise that no simple relationship of this sort exists in the time domain; for example, it is not possible to derive the output waveform by muliplying the input waveform by some time function representing the system itself. The main disadvantage of the frequency-domain approach, however, is that we very often want to know how a particular waveform is modified as it passes through a system. In this case, we must first find its spectrum (that is, take its Fourier transform), then multiply by the system's frequency response to get the output spectrum, and finally take the inverse Fourier transform to get the output time function. This can be both a lengthy and difficult procedure.

To illustrate these ideas, suppose a square wave signal forms the input to a linear processor having the frequency response

$$H(j\omega) = \frac{1}{1 + j\omega t_1}$$

This type of response may be readily achieved by a simple electrical network.[8] Its magnitude and phase characteristics are given by

$$|H(j\omega)| = \frac{1}{|1 + j\omega t_1|} = \frac{1}{\sqrt{[1 + (\omega t_1)^2]}}$$

$$\text{and } \phi(j\omega) = -\tan^{-1} \omega t_1.$$

These two characteristics, illustrated in figure 7.2, show that the processor transmits very low frequencies with negligible modification to amplitude and phase; but the transmission of higher frequencies is accompanied by both a reduction in magnitude and a phase lag, the latter approaching $\pi/2$ radians $(90°)$ as the frequency tends to infinity. The parameter t_1 is widely referred to as a 'time constant'.

he effect which such characteristics have on a square wave input depends very much on the latter's period. As we have seen in section 2.4.1, the Fourier series description of a repetitive square wave, symmetrical about $t = 0$ and having zero mean value, is of the form

$$\frac{4}{\pi} \{\cos \omega_1 t - \tfrac{1}{3} \cos 3\omega_1 t + \tfrac{1}{5} \cos 5\omega_1 t - \ldots \}.$$

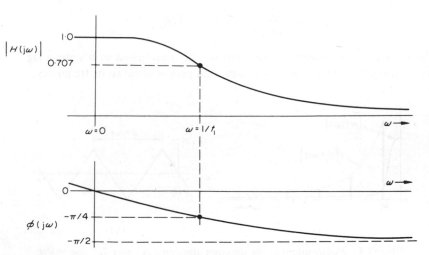

Figure 7.2 *Frequency-response characteristics of a simple linear processor*

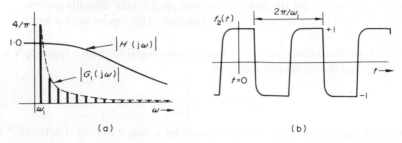

Figure 7.3   *(a) Spectral magnitude characteristics of an input square wave and system frequency response, and (b) the resulting output waveform*

where ω_1 is its fundamental frequency which has a period of $2\pi/\omega_1$ seconds. Let us consider two rather extreme cases: in the first, illustrated by figure 7.3(a), we assume that ω_1 is very small so that all the harmonics of substantial magnitude fall within that range of frequencies transmitted by the processor with little reduction in magnitude. Most components will arive at the output more or less unaltered in size, and as the phase characteristic of figure 7.2 suggests, with a phase lag approximately proportional to frequency—representing a pure time delay. Therefore the output waveform will be essentially similar to the input, but slightly delayed. The only harmonics which undergo substantial reduction are the high frequency ones: we must therefore expect the sudden transitions between the two levels of the input square wave not to be accurately reproduced at the output. A typical output waveform is shown in figure 7.3(b). In the second case, let us assume that ω_1 is much larger and that the fundamental component (and therefore all higher harmonics) undergoes very substantial reduction in passing through the system. This situation is illustrated by figure 7.4(a), in which the frequency axis has been much compressed. Now at sufficiently high frequency $\omega t_1 \gg 1$ and hence

$$H(j\omega) \to \frac{1}{j\omega t_1} \quad \text{and} \quad |H(j\omega)| \to \frac{1}{\omega t_1}$$

Therefore each component of the input will be transmitted with a phase lag approaching $\pi/2$ and an amplitude reduction proportional to its frequency.

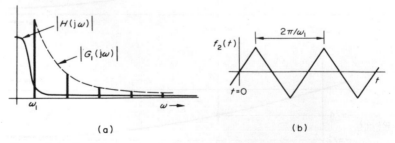

Figure 7.4   *Modifications to the situation illustrated in figure 7.3, due to the application of an input square wave of much shorter period*

Every cosine term in the input spectrum will therefore be converted to a sine, and its amplitude will also be modified, to give an output Fourier series of the form

$$f_2(t) = \frac{K}{\omega_1} \left\{ \sin \omega_1 t - \frac{1}{3} \cdot \frac{\sin 3\omega_1 t}{3} + \frac{1}{5} \cdot \frac{\sin 5\omega_1 t}{5} - \cdots \right\}$$

where K is a constant. The waveform represented by these frequency components is triangular in form and is shown in figure 7.4(b), which has both its axes suitably expanded (note that the amplitude of the triangular wave will be very small because each input component has been greatly reduced).

These two examples have been chosen to illustrate the ways in which modifications to spectral components can alter the shape of a repetitive waveform as it passes through a linear system. We have, of course, looked at two rather special cases; in general, a square wave input will not arrive at the output either more or less unchanged, or as a triangular wave. But the precise output waveform may always in principle be found by expressing the output spectrum in the form of a Fourier series, and evaluating it for a suitable set of instants in time.

So far, we have considered the case of a repetitive input signal by analysing changes in its various harmonic components as they pass through the system. The situation is somewhat more difficult to visualise, although similar in principle, with an aperiodic input signal having a continuous spectrum. Suppose, for example, an isolated decaying exponential waveform forms the input signal to the processor described above. The input is given by

$$\begin{aligned} f_1(t) &= e^{-\alpha t}, && 0 < t < \infty \\ &= 0, && -\infty < t < 0 \end{aligned}$$

Its spectrum may readily be evaluated by Fourier transformation as

$$G_1(j\omega) = \frac{1}{(\alpha + j\omega)} \qquad \text{(see section 3.3.4)}$$

Hence the output spectrum is

$$G_2(j\omega) = G_1(j\omega) \cdot H(j\omega) = \frac{1}{(\alpha + j\omega)} \cdot \frac{1}{(1 + j\omega t_1)}$$

In this case, inverse transformation to derive the corresponding time function is quite straightforward, since $G_2(j\omega)$ may be expressed in partial fraction form as

$$G_2(j\omega) = \frac{1}{1 - \alpha t_1} \left\{ \frac{1}{(\alpha + j\omega)} - \frac{1}{(\beta + j\omega)} \right\}$$

where $\beta = 1/t_1$. A term of the form $1/(\alpha + j\omega)$ represents a decaying exponential time function, and the output signal is therefore just equal to the difference between two such waveforms

$$f_2(t) = \frac{1}{1 - \alpha t_1}\left[\exp\left(-\alpha t\right) - \exp\left(-t/t_1\right)\right]$$

Typical input and output waveforms are illustrated in figure 7.5.

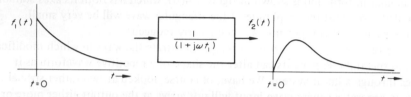

Figure 7.5   *Response of a linear system to a decaying exponential form of input waveform*

The above examples illustrate the general principles involved in estimating the form of output waveform for a given input and system frequency response: they also indicate how to specify the properties of a linear system for performing a particular signal processing task. For example, suppose we wish to convert a signal waveform $f_1(t)$ into a different waveform $f_2(t)$, using a linear signal processing operation. We take the Fourier transforms of $f_1(t)$ and $f_2(t)$; denoting these by $G_1(j\omega)$ and $G_2(j\omega)$, the required frequency response $H(j\omega)$ is given by

$$H(j\omega) = \frac{G_2(j\omega)}{G_1(j\omega)}$$

Of course, $G_2(j\omega)$ must not contain any frequency components absent from $G_1(j\omega)$ if $H(j\omega)$ is to describe a linear operation.

The methods outlined above for describing the relationship between input and output signals of a linear system involve analysis in terms of continuous sinusoidal waveforms: the input and output signals are represented as the sum of sets of continuous sinusoids (using the Fourier transform), and the linear system is represented by its effect on continuous sinusoids. This approach is often referred to as 'steady-state' sinusoidal analysis, because the sinusoidal functions are assumed to exist throughout all time. We now turn our attention to a somewhat more general frequency-domain description of a linear system, using the Laplace transform. As we have seen in section 3.4, the Laplace transform gives us a useful description of a signal waveform; not only does it allow us to derive a frequency-domain description of certain signals for which the Fourier integral fails to converge, but it also gives us the possibility of representing a signal by a set of s-plane poles and zeros. If we have found the Laplace transform of a signal we may, as a general rule, derive its (sinusoidal) frequency spectrum by substitution of $j\omega$ for the complex frequency variable s: the Laplace transform may therefore be considered a more general frequency-domain description than the Fourier transform.

Bearing in mind that a complex function $H(j\omega)$ may represent either a signal spectrum or a frequency response, it is not difficult to imagine that a Laplace transform $G(s)$ may also be used to describe either a signal or a system: when used to describe a system, it is generally referred to as a 'transfer function'. Just as the frequency response of a system defines its effect on continuous sinusoidal inputs, so the transfer function defines its effect on inputs of the form e^{st} (defined in the interval $0 < t < \infty$), where s may be real, imaginary or complex. Thus the transfer function defines the operation of a linear system for a rather wider range of input (and hence output) signal types than does the sinusoidal frequency response.

We now show that the transfer function of a linear signal processor is closely related to the form of differential equation relating its input and output in the time domain; as already stated, this equation is of the general form

$$c_n \cdot \frac{\mathrm{d}^n f_2(t)}{\mathrm{d}t^n} + c_{n-1} \cdot \frac{\mathrm{d}^{n-1} f_2(t)}{\mathrm{d}t^{n-1}} + \ldots c_1 \cdot \frac{\mathrm{d}f_2(t)}{\mathrm{d}t} + c_0 \cdot f_2(t) = f_1(t)$$

where $f_1(t)$ and $f_2(t)$ represent input and output signals respectively. Now the frequency-domain approach to the analysis of signals and systems involves transformation of time functions into functions of frequency. We could clearly take the Laplace (or Fourier) transform of the input signal $f_1(t)$ in the above equation, and use it to represent that input. We now show that it is also possible to derive the Laplace transform of the left-hand side of the equation, and use it to represent the linear system.

In section 3.4.4 some further properties of the Laplace transform were derived. One of these concerned the modification to a Laplace transform caused by differentiation of the corresponding time function. In the present context, if

$$f_2(t) \qquad \text{transforms to } G_2(s)$$

then

$$\frac{\mathrm{d}f_2(t)}{\mathrm{d}t} \qquad \text{transforms to } s \cdot G_2(s) - f_2(0),$$

$$\frac{\mathrm{d}^2 f_2(t)}{\mathrm{d}t^2} \qquad \text{transforms to } s^2 \cdot G_2(s) - s \cdot f_2(0) - \frac{\mathrm{d}f_2(0)}{\mathrm{d}t}$$

and so on. The constants $f_2(0)$ and $\mathrm{d}f_2(0)/\mathrm{d}t$ are the values of $f_2(t)$ and its first derivative immediately before $t = 0$, and represent in this case the state of the system immediately before the input signal $f_1(t)$ is applied. In other words, they allow for any situation where the output signal $f_2(t)$—or its derivatives—have non-zero values just before $f_1(t)$ is applied. This could arise if, for example, the system had not yet settled following the application of some previous input signal, and still had some energy stored in it. In most cases we assume the linear system to be in a quiescent state just before $f_1(t)$ is applied, and therefore that all such 'initial condition' constants are zero. Given this assumption, the nth derivative of $f_2(t)$ has a Laplace transform simply equal to s^n times that of $f_2(t)$ itself. Representing the Laplace transform of $f_1(t)$ by $G_1(s)$, we may therefore write down the transformed differential equation describing the system's performance as follows

$$c_n s^n . G_2(s) + c_{n-1} s^{n-1} . G_2(s) + \ldots c_1 s . G_2(s) + c_0 . G_2(s) = G_1(s)$$

or

$$G_2(s)[c_n s^n + c_{n-1} s^{n-1} + \ldots c_1 s + c_0] = G_1(s)$$

The great power of the frequency-domain approach is once again emphasised. For by taking Laplace transforms we have converted a differential equation in the time domain into an algebraic equation involving a polynomial in the complex frequency variable s; the latter equation is far simpler to manipulate.

We now define the transfer function $H(s)$ of the linear system as the ratio of the transformed input and output time functions. Hence

$$H(s) = \frac{G_2(s)}{G_1(s)} = \frac{1}{(c_n s^n + c_{n-1} s^{n-1} + \ldots c_1 s + c_0)}$$

In the most general case, the differential equation which describes the system in the time domain contains derivatives of $f_1(t)$ as well as of $f_2(t)$, and therefore gives rise to a function $H(s)$ containing both numerator and denominator polynomials in s. It should also be noted that

$$G_2(s) = G_1(s) . H(s)$$

This latter result is directly analogous to the relationship between input and output spectra and the system's frequency response

$$G_2(j\omega) = G_1(j\omega) . H(j\omega)$$

As an example of the use of the Laplace transform, suppose we have a linear system with a transfer function

$$H(s) = \frac{1}{(s + \alpha)}$$

to which a sinusoidal waveform $\sin \omega_1 t$ is applied as an input signal at $t = 0$. The Laplace transform of this input may be evaluated, or looked up in a table of transforms (such as that which follows the text); it is given by

$$G_1(s) = \omega_1 . \frac{1}{(s^2 + \omega_1^2)}$$

The Laplace transform of the output waveform (assuming the system is initially in a passive state) is therefore

$$G_2(s) = G_1(s) . H(s) = \frac{\omega_1}{(s + \alpha)(s^2 + \omega_1^2)}$$

Further reference to our table of transforms shows that the output waveform is given by

$$f_2(t) = \frac{\sin(\omega_1 t - \theta) + e^{-\alpha t} \sin \theta}{\sqrt{(\alpha^2 + \omega_1^2)}}$$

where

$$\theta = \tan^{-1}\left(\frac{\omega_1}{\alpha}\right).$$

This result, which is illustrated by figure 7.6, helps to emphasise an important point which may have troubled the reader with some experience of differential equations. The solution of a differential equation relating a linear system's response to a

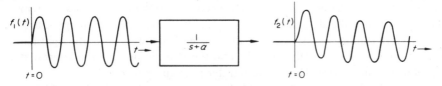

Figure 7.6   *Response of a linear system to a continuous sinusoidal input switched on at time t = 0*

sinusoidal input forcing function, applied to some instant such as $t = 0$, is quite often found in two separate parts. First is evaluated the form of the output or response to a truly continuous sinusoid, which is known as the 'particular integral': and this result is then modified by the so-called 'complementary function', to take account of the fact that the input is in fact applied suddenly at $t = 0$. It is important to realise that the Laplace transform method yields a solution which includes both the particular integral and the complementary function, and therefore embraces both the 'transient' and 'steady-state' parts of the response.[8] This is because the Laplace integral, by taking no account of what happens in the interval $-\infty < t < 0$, in effect assumes that all signals are switched on at time $t = 0$; switching effects are, in other words, an intrinsic part of the Laplace transform method and no separate account need be taken of them.

Returning to the subject of transfer functions, the most general form of such a function is

$$H(s) = \frac{(a_m s^m + a_{m-1} s^{m-1} + \ldots a_1 s + a_0)}{(c_n s^n + c_{n-1} s^{n-1} + \ldots c_1 s + c_0)}$$

The numerator and denominator polynomials in s may be factorised to give

$$H(s) = \frac{K \cdot (s - z_1)(s - z_2) \ldots (s - z_m)}{(s - p_1)(s - p_2) \ldots (s - p_n)}$$

where K is a constant. By direct analogy with the Laplace transform description of a signal, $z_1, z_2 \ldots z_m$ are called the zeros of $H(s)$ and $p_1, p_2 \ldots p_n$ the poles.

Since the coefficients $a_m, a_{m-1}, \ldots a_0$ and $c_n, c_{n-1}, \ldots c_0$ are always real, the poles and zeros of the transfer function are always either real or occur in complex conjugate pairs.[17] Furthermore, poles must all lie in the left-hand half of the s-plane if the transfer function is to represent a stable (and therefore usable) system. The reason for this may perhaps be most easily appreciated by considering a simple transfer function with a single real positive pole at $s = +\alpha$, so that

$$H(s) = \frac{1}{(s - \alpha)}$$

If any input signal is applied, the Laplace transform of the output will also contain a term $(s - \alpha)$ in its denominator: such a term represents a time function of exponential form which grows without limit as t increases. This means that even when the signal input is finite both in amplitude and in duration, the output is not, and the system is therefore unstable. (It is worth noting that a physical system which is active in the absence of a forcing function must have an energy source inside it). In summary, any system with transfer function poles in the right-half of the s-plane will be inherently unstable, and therefore of no practical interest as a signal processing device. This restriction does not apply to zeros.

A typical pole–zero plot for a stable linear system is shown in figure 7.7(a). From this plot we may infer the general form of the sinusoidal frequency response of the system in a way identical to that described in section 3.4.3 for signal spectra. A pole close to the imaginary axis in the s-plane gives rise to a hump, or 'resonant peak', in the frequency response at values of ω close to the pole's imaginary part; zeros give rise to corresponding troughs. A further point to note is that any practical system cannot have more zeros than poles in its transfer function: this is best explained by considering the form of the frequency response as $\omega \to \infty$. In general

$$H(j\omega) = H(s)\,|_{s=j\omega} = \frac{K(j\omega - z_1)(j\omega - z_2) \ldots (j\omega - z_m)}{(j\omega - p_1)(j\omega - p_2) \ldots (j\omega - p_n)}$$

and as $\omega \to \infty$

$$H(j\omega) \to K \cdot \frac{(j\omega)^m}{(j\omega)^n} \qquad \text{and} \qquad |H(j\omega)| \to K \cdot \frac{\omega^m}{\omega^n}$$

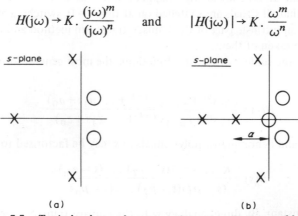

(a)   (b)

Figure 7.7   *Typical s-plane pole–zero configurations for (a) a stable linear system and (b) an output signal from that system*

But it is found that no practical system displays a frequency response with a magnitude which rises without limit as $\omega \to \infty$. Hence m must be less than or equal to n. Since the Laplace transform of the output of a system is the product of its input transform and the system transfer function, it follows that the pole–zero plot of the output signal is found by superimposing those of the input signal and of the transfer function. For example, figure 7.7(b) shows the pole–zero configuration of the output signal of a system described by the configuration of figure 7.7(a), when the input signal has a zero at $s = 0$ and a pole at $s = -\alpha$.

Although this discussion of transfer functions and pole–zero configurations has concentrated on the Laplace transform and the s-plane, the same general concepts are directly applicable to the z-transform and z-plane which, as we have seen in section 4.4, are particularly useful for the description of sampled-data signals. However it is only sensible to use the z-transform to describe the transfer function of a linear system when both input and output signals are in sampled form, as illustrated in figure 7.8. Thus when we speak of a linear system for the processing

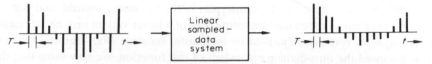

Figure 7.8   *Typical input and output signals of a linear sampled-data system*

of a sampled-data signal, we normally imply that the system output consists of a series of sample values, each one of which is coincident in time with the arrival of a sample at the input. The term 'linear' still implies the important frequency-preservation property noted above, in the sense that a sampled sinusoidal input waveform produces a sampled sinusoidal output at the same frequency (but having, in general, different amplitude and phase). The transfer function of such a system is given by

$$H(z) = \frac{Y(z)}{X(z)}$$

where $Y(z)$ and $X(z)$ are the z-transforms of the output and input sampled-data signals respectively. Now $z = e^{sT}$, where T is the sampling period of the system, and we may obtain the sinusoidal frequency response of the system by substituting $j\omega$ for s, giving

$$H(j\omega) = \frac{Y(e^{j\omega T})}{X(e^{j\omega T})}$$

We have already noted in chapter 4 that the frequency spectrum of a sampled-data signal is essentially repetitive in form: therefore we must expect that the frequency response of the type of sampled-data system shown in figure 7.7 will also be repetitive, and this is indeed the case. However, rather than deal at length with the frequency-domain properties of sampled-data systems at this stage, it is almost certainly more helpful to keep only these general points in mind, and to pass on to a discussion of the time-domain relationships between signals and linear systems. The practical effects of the repetitive frequency response of a sampled-data system will become clearer during our discussion of signal sampling in chapter 8 and of sampled-data (that is, digital) filters in chapter 9.

7.2.3 The time-domain approach

As we have seen, frequency functions representing an input signal and a linear processor may be multiplied together to yield a frequency function representing the output. This basic fact is responsible for the popularity of frequency-domain analysis. On the other hand, we are often interested in the modifications to a signal waveform caused by processing, and the transformation of time functions into their frequency-domain equivalents must at times seem a longwinded way to the answer. We shall show in this section that it is possible to work entirely with time functions, even if the manipulation involved is by no means as simple as multiplication.

Every function of time has its counterpart in the frequency domain, and vice versa. We have already seen how the properties of a linear system may be represented by its frequency response $H(j\omega)$—or by its transfer function $H(s)$—but we have not so far discussed the time-domain equivalent of this function. We now show that the corresponding time function is in fact the response of the system to a short, isolated, impulsive input, and that it gives the clue to the time-domain approach.

The idea of a Dirac pulse was introduced in section 4.2, where its spectrum was shown to be

$$G(j\omega) = 1$$

Furthermore, reference to the table at the end of the text shows that its Laplace transform is also unity. Suppose such a pulse forms the input signal to a linear system having a transfer function $H(s)$. The Laplace transform of the output is simply

$$G_2(s) = H(s) \cdot G_1(s) = H(s) \cdot 1 = H(s)$$

The corresponding time function is clearly the response of the system to the Dirac pulse, and is called its 'impulse response'. The impulse response and transfer function are therefore related as a Laplace (or Fourier) transform pair. The impulse response is just as valid a description of the system's properties as its transfer function; the two measures give identical information about the system, although in rather different forms.

In this book we are concerned with two main classes of linear system: what are generally referred to as continuous systems, which handle continuous input waveforms and produce continuous outputs; and sampled-data systems, which handle sampled input signals and produce sampled outputs. The impulse response of the first type of system represents its output due to an extremely narrow input pulse; that of the latter may conveniently be thought of as its response to a single unit-height input

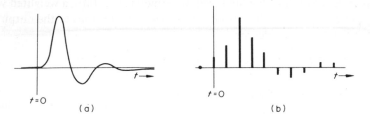

Figure 7.9 *Typical impulse responses of (a) a continuous system and (b) a sampled-data system*

sample. Typical examples of these two types of impulse response are illustrated in figure 7.9, adopting the usual convention that the input impulse is applied at time $t = 0$ (no physical system can respond before an input is delivered, which means that a realisable impulse response function must be zero before $t = 0$).

In principle, the time-domain approach is quite simple: we consider an input signal to be composed of a succession of impulse functions, each of which generates a weighted version of the impulse response at the output of the linear system. The output signal is then found by the superposition of all such responses. This concept is, in a striking sense, the exact antithesis of the frequency-domain approach. For instead of considering the input to be made up of extended time functions (for example, continuous sinusoids), each of which has an infinitely narrow spectrum, we now think of it as being composed of extremely narrow time-pulses each of which has an infinitely broad spectrum. There are two common situations to which this time-domain approach is naturally suited; the first arises when a linear system is disturbed by an input of restricted duration, such as an isolated pulse or transient waveform; the second, when the input is in the form of a whole series of narrow pulses separated from one another in time—in other words, a sampled-data signal.

The way in which an output signal may be derived by use of the impulse response is perhaps most easily appreciated in the case of a sampled-data system. Suppose, for example, we have an impulse response consisting of sample values $a_0, a_1, a_2, \ldots a_n$ and an input signal which begins at some reference time $t = 0$ and has successive values $x_0, x_1, x_2, x_3 \ldots$, as shown in figure 7.10. Let us now estimate, say, the output sample value y_3 which coincides with the arrival of x_3

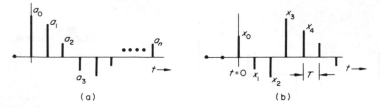

Figure 7.10 *(a) The impulse response of a sampled-data system, and (b) an input signal applied to it*

at the input. Since each of the input samples generates, in turn, a weighted version of the impulse response at the output, successive contributions to the output due to the first few input samples are

at $t = 0$	at $t = T$	at $t = 2T$	at $t = 3T$	at $t = 4T$	
$\downarrow$	$\downarrow$	$\downarrow$	$\downarrow$	$\downarrow$	
$x_0 a_0$ +	$x_0 a_1$ +	$x_0 a_2$ +	$x_0 a_3$ +	$x_0 a_4$	+ ..., due to x_0
	$x_1 a_0$ +	$x_1 a_1$ +	$x_1 a_2$ +	$x_1 a_3$	+ ..., due to x_1
		$x_2 a_0$ +	$x_2 a_1$ +	$x_2 a_2$	+ ..., due to x_1
			$x_3 a_0$ +	$x_3 a_1$	+ ..., due to x_3
				$x_4 a_0$	+ ..., due to x_4

Therefore the total output at the instant $(t = 3T)$ corresponding to the arrival of x_3 at the input is

$$y_3 = x_0 a_3 + x_1 a_2 + x_2 a_1 + x_3 a_0$$

In other words, each input sample contributes to the output at a particular sampling instant in accordance with (a) its value and (b) elapsed time. There is a very convenient graphical method of obtaining the above result, which involves laying the impulse response out backwards from the instant for which the output is to be calculated, and summing all cross-products of coincident terms in impulse response and input signal. This method is illustrated by figure 7.11, which shows that the sum of all finite cross-products so obtained is indeed, as before

$$x_0 a_3 + x_1 a_2 + x_2 a_1 + x_3 a_0$$

This result emphasises that the output of a linear sampled-data system at any sampling instant is equal to the sum of a number of input samples, weighted by a

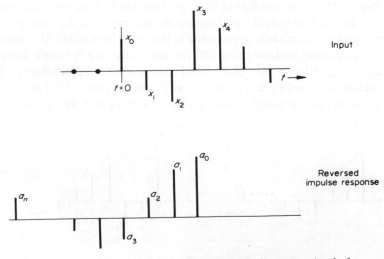

Figure 7.11   *A graphical method of evaluating the output signal of a sampled-data system*

set of constants which define the system's impulse response. The above procedure for calculating the output is called 'convolution', and, as we have shown, it involves reversal of the impulse response, cross-multiplication, and summation. Convolution is no different in principle when both input signal and impulse response are continuous time functions, even if it is rather more difficult to visualise. In such a case the input is in effect considered to be made up from a very large number of contiguous Dirac impulses, as shown in figure 7.12. Each impulse has a standard width, and a height (and hence area) given by the value of $f_1(t)$ at the instant considered.

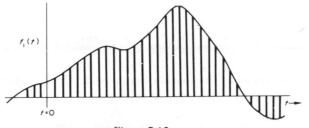

Figure 7.12

Although the above introduction to the idea of convolution is essentially intuitive, the convolution procedure may be expressed in precise mathematical form. The convolution theorem states that if three time functions $f_1(t)$, $I(t)$, and $f_2(t)$ have Fourier transforms $G_1(j\omega)$, $H(j\omega)$, and $G_2(j\omega)$, and if

$$G_2(j\omega) = G_1(j\omega) . H(j\omega)$$

then the time functions satisfy the relationship

$$f_2(t) = \int_{-\infty}^{\infty} f_1(T') . I(t - T') . dT'$$

where T' is an auxiliary time variable. This relationship, as we will show later, is the mathematical counterpart of the graphical procedure described above, and is known as the convolution integral. It is interesting that the convolution theorem says nothing about signals and linear systems: it merely states that multiplication of two frequency functions $G_1(j\omega)$ and $H(j\omega)$ is equivalent to convolution of their corresponding time functions. The validity of this result may be demonstrated without great difficulty. Since $G_2(j\omega)$ is the Fourier transform of $f_2(t)$, the transform of the right-hand side of the convolution integral equation should equal $G_1(j\omega) . H(j\omega)$. This transform is

$$\int_{-\infty}^{\infty} \left\{ \int_{-\infty}^{\infty} f_1(T') . I(t - T') \, dT' \right\} e^{-j\omega t} . dt$$

$$= \int_{-\infty}^{\infty} \left\{ \int_{-\infty}^{\infty} I(t - T') \, e^{-j\omega t} \, dt \right\} f_1(T') . dT'.$$

It is legitimate to change the order of integration if the integrals are convergent. Now the quantity in brackets is the Fourier transform of $I(t - T')$, which is the same waveform as $I(t)$ apart from a delay of T' seconds. We have already seen (section 4.2) that delaying a waveform by T' seconds is equivalent to multiplying its spectrum by $e^{-j\omega T'}$; therefore the transform of $I(t - T')$ is $[H(j\omega) . e^{-j\omega T'}]$. Our expression now becomes

$$\int_{-\infty}^{\infty} \left\{ H(j\omega) . e^{-j\omega T'} \right\} f_1(T') . \mathrm{d}T'$$

$$= H(j\omega) \int_{-\infty}^{\infty} f_1(T') . e^{-j\omega T'} \, \mathrm{d}T' = H(j\omega) . G_1(j\omega)$$

which is the result stated by the convolution theorem. The convolution integral may be expressed slightly differently, by writing $(t - T')$ as T''. Then

$$T' = t - T'' \qquad \text{and} \qquad \mathrm{d}T' = -\mathrm{d}T''$$

Furthermore

$$\text{if } T' = \infty, \qquad T'' = -\infty; \qquad \text{and if} \qquad T' = -\infty, \qquad T'' = \infty$$

Substitution yields the form

$$f_2(t) = \int_{\infty}^{-\infty} f_1(t - T'') . I(T'') . (-\mathrm{d}T'') = \int_{-\infty}^{\infty} f_1(t - T'') . I(T'') . \mathrm{d}T''$$

Let us now relate this form of the convolution integral to the graphical procedure used earlier. The first thing to note is that $I(t)$ has been defined as the transform of $H(j\omega)$ and therefore represents the impulse response of the system. Suppose we are considering the value of the output signal at some instant t_1, and we interpret the auxiliary time variable T'' (which appears in our second version of the convolution integral) as time measured back into the past from t_1. The value of the system output at time t_1 is given by

$$f_2(t_1) = \int_{-\infty}^{\infty} f_1(t_1 - T'') . I(T'') . \mathrm{d}T''$$

Now $I(T'')$ traces out the same function for increasing T'' as $I(t)$ traces out for increasing t. Hence $I(T'')$ represents the impulse response laid out backwards in time from the instant t_1. Furthermore

$$f_1(t_1 - T'') = f_1(0) \qquad \text{when} \qquad T'' = t_1, \text{ which corresponds to the instant } t = 0.$$

$$= f_1(t_1) \qquad \text{when} \qquad T'' = 0, \text{ which corresponds to the instant } t = t_1.$$

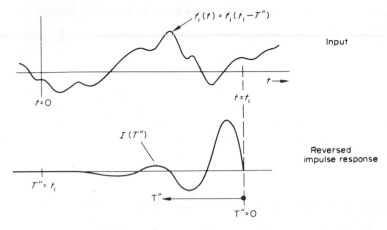

$f_1(t) = f_1(t_1 - T'')$

Input

$t = 0$

$t \longrightarrow$

$t = t_1$

$I(T'')$

Reversed
impulse response

$T'' = t_1$

$T'' \longleftarrow$

$T'' = 0$

Figure 7.13   *Graphical interpretation of the convolution integral for the case
of a continuous input signal applied to a continuous system*

Hence $f_1(t_1 - T'')$ is simply the input signal $f_1(t)$. Therefore the convolution
integral implies, once again, laying out the impulse response backwards from the
instant t_1 and multiplying it with the input signal waveform; this cross-product is
integrated over all time—or, in practice, over that part of the time-scale for which
it is nonzero. In the case of discrete (that is, sampled-data) signals, the integration
becomes a simple summation of terms, as we saw earlier. The graphical interpretation of the convolution integral for the case of a typical continuous system
and input signal is illustrated in figure 7.13. It should be noted that the limits of
integration of the convolution integral are quite often given as 0 and ∞. This need
cause no concern: whenever we deal with a physically realisable system, the value of
$I(T'')$ for $T'' < 0$ must be zero, because the impulse response cannot start before the
impulse which causes it. Therefore the contribution to the integral in the region
$-\infty < T'' < 0$ must also be zero, and limits of 0 and ∞ are adequate.

   It was earlier stated that the time-domain approach is particularly suitable when
we are dealing either with a sampled-data system, or with a continuous system
disturbed by a transient (that is, time-limited) input signal. The first case has
already been discussed, and we now turn our attention to an example of the
second. Figure 7.14 shows an input signal applied to a linear system having an
impulse response of the form

$$I(t) = \frac{1}{t'} e^{-t/t'}, \qquad \text{for} \qquad t > 0.$$

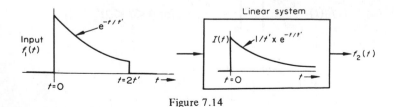

Figure 7.14

(The corresponding frequency response may easily be shown by Fourier transformation to be

$$H(j\omega) = \frac{1}{(1 + j\omega t')}$$

This form of response was discussed at some length in the previous section and illustrated in figure 7.2.) Considerable care must be taken in specifying the limits of integration in this case, because the input signal is switched abruptly on and off. The limits are best decided by once again sketching the reversed impulse response below the input waveform, as shown in figure 7.15(a). From this we see that when

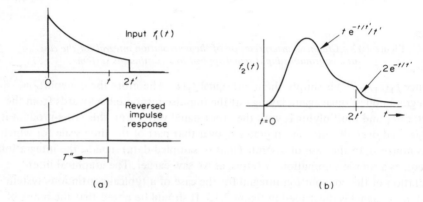

(a)   (b)

Figure 7.15

the instant for which the output is to be evaluated is less than $2t'$, the limits are $T'' = 0$ and $T'' = t$; otherwise they are $T'' = t - 2t'$ and $T'' = t$. The indefinite convolution integral is

$$f_2(t) = \int f_1(t - T'') . I(T'') . dT''$$

$$= \int e^{-(t-T'')/t'} . \frac{1}{t'} . e^{-T''/t'} . dT''$$

$$= \frac{1}{t'} . e^{-t/t'} . \int dT'' = \frac{T''}{t'} e^{-t/t'}$$

Inserting the limits, we have

$$f_2(t) = \frac{e^{-t/t'}}{t'} \left[T'' \right]_0^t = \frac{t \, e^{-t/t'}}{t'}, \text{ for } 0 < t < 2t'$$

and

$$f_2(t) = \frac{e^{-t/t'}}{t'} \left[T'' \right]_{t-2t'}^t = 2e^{-t/t'}, \text{ for } t > 2t'.$$

The form of the output is shown in figure 7.15(b). Although this result could have been obtained by using frequency-domain methods, the convolution integral is simple to evaluate and provides a very convenient method of solution.

Before leaving this introduction to the time-domain approach, several points of general interest are worth making. Firstly, the convolution procedure involves, in effect, multiplying or weighting the input signal $f_1(t)$ by the system's impulse response, and then integrating. For this reason the impulse response of a linear system is sometimes called its 'weighting function'. The weighting function tells us how the previous history of the input signal affects the present value of the output: the more extended in time the weighting function, the more extended in time is that portion of the input's past history which affects the output at a particular instant. When previous values of an input affect the present value of an output, the system may be said to possess a memory; this accounts for the occasional use of a further name for the impulse response—the 'memory function'. The only form of impulse response which does not involve memory is one which itself consists of an impulse, which implies that the system responds equally to all frequencies. Hence any frequency-dependent system must involve memory. It is also interesting to note that the duration of the impulse response defines the time for which the system output continues after the input signal has been removed (this is well illustrated by the example of figures 7.14 and 7.15): similarly, it indicates the duration of the switching transient which occurs when a sinusoidal input signal of any frequency (including zero frequency—that is, a steady level) is suddenly switched on (see, for example, figure 7.6). For only when the input signal has been turned on for at least as long as the duration of the impulse response does the output reach its so-called 'steady-state' condition. An initial switching transient is therefore to be viewed as an entirely necessary result of using any linear system whose performance is frequency-dependent.

Since a Dirac impulse has the spectrum

$$G(j\omega) = 1$$

all frequencies are equally represented in it. Therefore the application of a unit impulse as input signal to a linear system is tantamount to disturbing it, simultaneously, with cosine waves of all possible frequencies. Of these, some are preferentially transmitted by the system, and this fact is reflected by the precise form of the impulse response: in other words, the impulse response waveform is made up from those frequency components which are most effectively transmitted.

7.2.4 Examples of linear processes

The purpose of this section is to outline some basic linear processes which find widespread practical applications. All the processes mentioned may in practice be achieved, or at least approximated, by electrical or electronic circuits, or by suitable programming of a digital computer which is fed with sampled input-signal data.

Among the most common signal processing requirements are amplification and attenuation. Ideally, both produce a signal $f_2(t)$, related at every instant of time to the original $f_1(t)$ by a constant. Hence

$$f_2(t) = A \cdot f_1(t)$$

where for amplification $A > 1$, and for attenuation $A < 1$. Apart from a change of scale, the waveforms $f_2(t)$ and $f_1(t)$ are identical, as are their spectral characteristics. This implies that an amplifier or attenuator should have no 'memory', and that its present output should depend only on its present input. Ideally, its impulse response should therefore itself be a (weighted) impulse.

Mixing and filtering are also very common linear signal processes, and are essentially complementary to one another. By mixing is meant the addition of two or more signal waveforms (with or without amplification). Filtering, on the other hand, implies the separation of one signal from one or more others, normally on the basis of differences in their frequency spectra, and is perhaps the most important of all signal processing operations. The design of filters is crucial to many types of electronic equipment including radio and television receivers, which have to select one transmitted signal in the presence of many others, and is covered much more fully in chapter 9. Typical mixing and filtering processes are illustrated in figure 7.16. It should be noted that the mixer produces an output which at every instant equals the sum of two (or more) input waveforms, regardless of their shapes. An ideal mixer is therefore essentially a 'wideband' device, working equally well with any form of input signal (and hence with any form of input spectrum). On the other hand a filter is frequency-selective: in most cases, a filter 'accepts' one band of frequencies which it delivers to the output, and rejects the rest; the rejected components of the input signal are not usually available as a second output (unlike the scheme indicated in figure 7.16). The most general use of the term 'filter' merely implies some sort of frequency-selective linear system, and embraces a very wide variety of practical signal processors.

A common application of mixers and filters is the setting of a signal's average level. For example, a signal with a finite average level (that is, a signal having a zero-frequency component) may have this average level removed by a filter, a process known in electrical engineering as 'd.c. removal'. The opposite process, that of

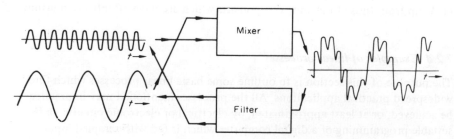

Figure 7.16 *Mixing and filtering*

adding an average level to a signal, is known as 'd.c. restoration' and may be achieved by using a mixer. A slight variation on 'd.c. restoration' is 'clamping', in which either the peak or minimum value of a repetitive waveform is clamped to a certain defined level, normally zero: clamping may be achieved either by linear mixing, or by the use of a nonlinear processor. D.C. removal, d.c. restoration, and clamping are illustrated by figure 7.17.

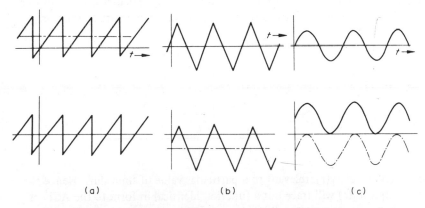

Figure 7.17 *(a) d.c. removal, (b) d.c. restoration, and (c) two types of clamping. In each case the top diagram represents the signal prior to processing*

7.3   Convolution, correlation, and filtering

The form of the convolution integral

$$f_2(t) = \int_{-\infty}^{\infty} f_1(T') . I(t - T') . dT'$$

is very similar to that defining the 'finite' version of the cross-correlation function relating two variables $f_1(t)$ and $f_2(t)$

$$r_{xy}(\tau) = \int_{-\infty}^{\infty} f_1(t) . f_2(t + \tau) . dt$$

In the first case the integral implies multiplication of f_1 with a time-reversed version of I, followed by integration over all time to give the value of f_2 relevant to a particular instant t; in the second case the product of f_1 and a time-shifted version of f_2 is integrated to give the value of r_{xy} relevant to a particular value of τ. The only substantial difference is that the convolution integral, unlike the correlation integral, requires a reversal of one of the time functions. We now explore the relationship between convolution and correlation rather more carefully, and attempt an explanation for the essential similarity of these two important processes.

Their relationship is in fact very effectively demonstrated by the idea of a 'matched filter'. Suppose a signal waveform $f_1(t)$ is applied as input to a linear system which has an impulse response equal to a time-reversed version of $f_1(t)$, as illustrated by figure 7.18. Ignoring, for the present, the question whether such a linear system may actually be constructed, let us consider the form of the resulting output waveform $f_2(t)$. Recalling the graphical interpretation of the convolution integral, we must time-reverse the impulse response, multiply it by the input signal, and integrate to find the output at a particular instant: in this case, such a procedure is identical to that for obtaining the autocorrelation

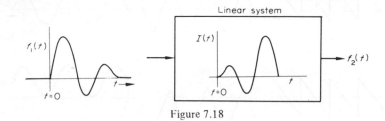

Figure 7.18

function (ACF) of $f_1(t)$, relevant to a particular value of time-shift. Hence the system output $f_2(t)$ will trace out a function identical in form to the ACF of $f_1(t)$; such a linear system is called a 'matched filter', since its characteristics are in a special sense 'matched' to those of its input waveform. Convolution of the matched filter's impulse response with the input signal is equivalent to auto-correlating the input. In a similar fashion, we could in principle generate a time-function identical in shape to the cross-correlation function of two different waveforms $f_1(t)$ and $f_2(t)$, by feeding $f_1(t)$ into a linear system whose impulse response was equal to a time-reversed version of $f_2(t)$.

It is rewarding to consider the frequency response of the filter 'matched' to a signal $f_1(t)$. As argued above, its impulse response is given by

$$I(t) = f_1(-t)$$

Its frequency response is therefore

$$H(j\omega) = \int_{-\infty}^{\infty} I(t) . e^{-j\omega t} . dt = \int_{-\infty}^{\infty} f_1(-t) . e^{-j\omega t} . dt$$

Let us write t' in place of $(-t)$, giving the alternative form

$$H(j\omega) = \int_{-\infty}^{\infty} f_1(t') . e^{j\omega t'} . dt'$$

But the spectrum of the input signal $f_1(t)$ is given by

$$G_1(j\omega) = \int_{-\infty}^{\infty} f_1(t) . e^{-j\omega t} . dt$$

and therefore $H(j\omega) = G_1 [j(-\omega)]$.

Now the spectrum $G_1(j\omega)$ of any real time function $f_1(t)$ may always be written in the form[16]

$$G_1(j\omega) = a(\omega) + jb(\omega)$$

where $a(\omega)$ is an even function of ω representing cosine components, and $b(\omega)$ is an odd function of ω representing sines. Hence

$$G_1[j(-\omega)] = a(-\omega) + jb(-\omega)$$
$$= a(\omega) - jb(\omega) = G_1{}^*(j\omega)$$

where the asterisk denotes the complex conjugate. This result shows that the frequency response of a matched filter is equal to the complex conjugate of the input signal spectrum to which it is matched. The output signal's spectrum is clearly given by

$$G_2(j\omega) = G_1(j\omega) \cdot H(j\omega) = G_1(j\omega) \cdot G_1^*(j\omega)$$
$$= |G_1(j\omega)|^2 = P_{xx}(\omega)$$

where $P_{xx}(\omega)$ is the energy spectral density of the input signal. As we have seen in section 5.4.3, the power or energy spectrum and autocorrelation function of a signal are related as a Fourier transform pair: hence, once again, we infer that the output waveform from the matched filter has the form of the ACF of the input.

Since any frequency-selective system or processor may be termed a 'filter', we see that the concepts of convolution, correlation, and filtering are closely interwoven. Correlation as a time-domain operation must therefore be viewed as equivalent to a form of frequency-domain filtering; the temptation to regard it as some rather special form of signal operation, quite divorced from the more familiar concept of filtering, should be avoided.

Although the idea of a matched filter is extremely valuable for clarifying the relationships between convolution, correlation, and filtering, it must be understood that such a filter is not generally realisable[4] when the signal $f_1(t)$ (and hence the impulse response $I(t)$) is a continuous function of time. On the other hand it is quite possible to realise a matched filter for a finite length sampled-data signal, either by suitable programming of a digital computer or by the use of special purpose electronic circuits. Such filters have special relevance for detecting signals in the presence of random disturbances, and are further discussed in chapter 10.

7.4 The processing of random signals

When a random waveform passes through a linear signal processor its time-averaged properties—as measured by its amplitude distribution, autocorrelation function, or power spectrum—are usually modified. Whether the random waveform represents a useful signal or an unwanted interference, it is clearly desirable to quantify the effects which a linear signal processing operation has upon it.

We have seen that a linear system is normally characterised either by its frequency response (or transfer function), or by its impulse response, and that these functions are equivalent descriptions of the system in frequency and time domain respectively. Not surprisingly, therefore, it is a relatively straightforward matter to discuss the effect which a linear system has on the frequency and time domain measures of a random signal. On the other hand, the effects of signal processing on such properties of a random waveform as its amplitude distribution or central moments are not generally easy to assess, because these properties are not simply related to its time and frequency domain structure.

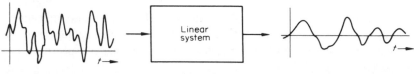

Figure 7.19

Figure 7.19 illustrates the type of situation we are now considering: a random waveform is applied as input to a linear system, and gives rise to a modified, but still random, output. Let us begin by considering the effect of the system on the spectral properties of the random signal. As we have seen in section 5.4.3, the power spectral density defines the average power of the various frequency components in a signal, but ignores their relative phases. Hence the power spectral density of the output from a linear system cannot depend upon the phase response of the system, but only upon its magnitude response. Suppose, for example, we apply a sinusoidal wave $A \sin \omega t$ to the input of a linear system and obtain an output $AB \sin (\omega t + \phi)$. The average input power is $A^2/2$ and the average output power is $(AB)^2/2$. Therefore a system having a response magnitude B at a frequency ω modifies the power of a component at that frequency by a factor B^2. More generally, if the input signal contains many frequency components and has a power spectral density $P_{xx}(\omega)$, then the output power spectral density will be

$$P_{yy}(\omega) = P_{xx}(\omega) . |H(j\omega)|^2 = P_{xx}(\omega) . H(j\omega) . H^*(j\omega)$$

where $H(j\omega)$ is the frequency response of the system, and the asterisk denotes the complex conjugate. Each input component is multiplied by the square of the response magnitude at the relevant frequency.

In section 7.2.3 we introduced the general principle that multiplication of two frequency functions is equivalent to convolution of their respective time functions. In the present context we note that the output power spectrum from a linear system is found by multiplying the input power spectrum by the system function $|H(j\omega)|^2$; since the power spectrum and autocorrelation function of a signal are related as a Fourier transform pair, it therefore follows that the ACF of the output signal from a linear system may be found by convoluting that of its input with the time function corresponding to $|H(j\omega)|^2$. This latter time function is in fact equal to the ACF of the system's impulse response, and is denoted by $J(\tau)$ in figure 7.20, where the above relationships are clarified.

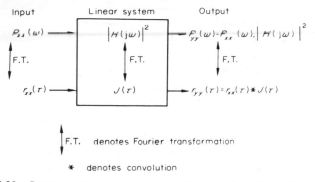

Input   Linear system   Output

$P_{xx}(\omega)$ —— $\left|H(j\omega)\right|^2$ —— $P_{yy}(\omega)=P_{xx}(\omega).\left|H(j\omega)\right|^2$

F.T.   F.T.   F.T.

$r_{xx}(\tau)$ —— $J(\tau)$ —— $r_{yy}(\tau)=r_{xx}(\tau)*J(\tau)$

F.T.   denotes Fourier transformation

*   denotes convolution

Figure 7.20   *Power spectrum and autocorrelation function relationships at input and output of a linear system*

Although our major interest in this chapter is to describe how signals are modified by linear processing, there is another side to the same coin: knowledge of the properties of input and output signals allows us to infer those of an unknown linear system. For example (and as already pointed out in section 7.2.2), if we know the input signal spectrum $G_1(j\omega)$ and the output signal spectrum $G_2(j\omega)$ of a system, we may express its frequency response quite simply as

$$H(j\omega) = \frac{G_2(j\omega)}{G_1(j\omega)}$$

Similarly if a random signal with a power spectrum $P_{xx}(\omega)$ is applied to a linear system and we measure the output power spectrum $P_{yy}(\omega)$, we are in a position to define the system's response magnitude $|H(j\omega)|$, since

$$|H(j\omega)|^2 = \frac{P_{yy}(\omega)}{P_{xx}(\omega)}$$

Of course, any practical estimates of $P_{xx}(\omega)$ and $P_{yy}(\omega)$ based on finite portions of random input and output signals will contain sampling errors, so that the estimate of $|H(j\omega)|$ will only be approximate.

Unfortunately, comparison of input and output power spectra fails to reveal any information about the way in which a system modifies the phases of various frequency components applied to its input, and it is interesting to consider whether any other type of comparison of random input and output signals might be used to define the system's response in phase as well as in magnitude. In chapter 6 we discussed signal comparison using the cross spectral density and cross-correlation functions, and saw that these measures reflect not only magnitudes but also relative phases of common frequency components in two signals $f_1(t)$ and $f_2(t)$. If we now consider $f_1(t)$ and $f_2(t)$ to represent random input and output waveforms of a linear system, it is clear that any measure which indicates relative phases of their various common components must indicate phase changes introduced by the system itself. This is the clue to a complete definition of a linear system by examination of the properties of its random input and output. In order to examine

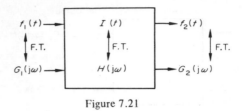

Figure 7.21

this question rather more carefully, suppose that the various signal and system descriptions in time and frequency domains are as illustrated in figure 7.21. Recalling that correlation and convolution are similar time-domain operations apart from reversal of one of the time functions, we may write down the following list of equivalent operations

(a) cross-correlation of $f_1(t)$ and $f_2(t)$,
(b) convolution of $f_1(-t)$ and $f_2(t)$,
(c) multiplication of $G_1(-j\omega)$ and $G_2(j\omega)$.

But $G_1(-j\omega) = G_1^*(j\omega)$ when $f_1(t)$ is a real time function: furthermore, $G_2(j\omega) = G_1(j\omega).H(j\omega)$, so that

$$G_1(-j\omega).G_2(j\omega) = G_1^*(j\omega) . G_1(j\omega) . H(j\omega)$$

$$= |G_1(j\omega)|^2 . H(j\omega)$$

Now $|G_1(j\omega)|^2$ is the power spectral density of the input signal, which is the Fourier transform of its autocorrelation function $r_{xx}(\tau)$; therefore operation (c) above is equivalent to

(d) multiplication of $|G_1(j\omega)|^2$ and $H(j\omega)$,

and hence to:

(e) convolution of $r_{xx}(\tau)$ and $I(t)$.

This last result is most interesting: we started by assuming cross-correlation of the input and output waveforms of the system, and result (e) shows that this is equivalent to convoluting the impulse response of the linear system with the ACF of its input. Formally the relationship may be stated by the convolution integral

$$r_{xy}(\tau) = \int_{-\infty}^{\infty} r_{xx}(\tau - T).I(T).dT$$

If we know the input ACF and evaluate the cross-correlation function relating the system input $f_1(t)$ and its output $f_2(t)$, we may therefore evaluate the impulse response of the system. The above formula becomes particularly useful when the input $f_1(t)$ is a very wideband random signal, which has an ACF approximating a Dirac pulse at $\tau = 0$ (see section 5.5.3 and figure 5.15). In this case we have

$$r_{xy}(\tau) = \int_{-\infty}^{\infty} \delta(\tau - T).I(T).dT$$

which, by the sifting property of the Dirac pulse, reduces directly to

$$r_{xy}(\tau) = I(\tau)$$

In other words the input-output cross-correlation function has the same shape as
the impulse response of the linear system. It is at first rather hard to believe this
result, since we are apparently deriving complete and deterministic information
about a system by comparing its random input and output waveforms. The essential
point, however, is that our estimate of $I(t)$ will involve sampling errors and will only
be accurate if we cross-correlate very long portions of the input and output random
signals. Figure 7.22(a) shows a typical case. A wideband random signal is applied

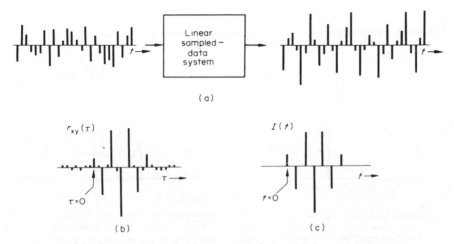

(a)

(b)

(c)

Figure 7.22   *System exploration by cross-correlation. (a) shows typical portions
of input and output signals of a sampled data system; (b) shows a practical
estimate of their cross-correlation function, whereas (c) is the actual impulse
response of the system*

to a linear sampled-data system, and is cross-correlated with its output. Using typical
input and output signal portions of length 1000 samples, the cross-correlation
function is as shown in figure 7.22(b), which compares favourably but not exactly
with the actual system impulse response shown in part (c) of the figure.

This technique of system identification by input-output cross-correlation is of
considerable practical interest. Since the cross-correlation function takes the same
form as the system's impulse response only when the input random signal is
wideband (in practice this means that the input power spectrum must be constant
or 'flat' over the full range of frequencies significantly transmitted by the system),
the value of wideband random or pseudo-random signals for system identification
is emphasised. It should also be noted that a random disturbance is quite often
present at the input (and hence output) of a linear system anyway; normally such
'noise' is merely a nuisance, but it can sometimes allow the above technique to be
put to good use. The characteristics of electronic circuits, chemical process plants,
and physiological systems have all been explored by input-output cross-correlation.

Although knowledge of the spectral properties of a linear system allows us to define its effect on the power spectrum (or autocorrelation function) of a random signal, and same is unfortunately not true of the signal's amplitude distribution. The relationship between input and output signal amplitude distributions is not generally a simple one; neither is it unique. To take an example, suppose that two random binary signals with the same amplitude distribution form alternative inputs to a typical linear system, as shown in figure 7.23: the amplitude distribution of the output waveform is quite different in the two cases, as the lower part of the figure shows. Indeed a given input amplitude distribution can give rise to any

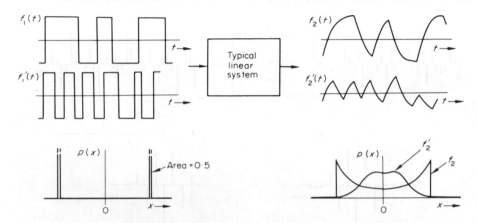

Figure 7.23   *Portions of two alternative random inputs and outputs of a linear system, and, below, their corresponding amplitude distributions: the amplitude distributions of the two inputs are identical, but those of the outputs are quite different*

number of output distributions, and the precise relationship between the two may not easily be described. There are in fact only two points of contact between the description of a random signal by its amplitude distribution and by its spectrum— namely the mean value, and the average power. As we have seen in section 5.3, the mean, or average, value of a signal described by an amplitude probability density function $p(y)$ is

$$\bar{y} = \int_{-\infty}^{\infty} p(y).y.dy$$

The mean value of a random output signal from a linear system may, of course, be found by multiplying the mean of the input by the value of the system's response relevant to zero frequency. Secondly, the average power or mean square value of a signal is given by

$$\overline{y^2} = \int_{-\infty}^{\infty} p(y).y^2.dt$$

and is also equal to the value of its autocorrelation function at $\tau = 0$. Further, since

$$r_{xx}(\tau) = \frac{1}{2\pi} \int_{-\infty}^{\infty} P_{xx}(\omega) \cdot e^{j\omega\tau} \cdot d\omega$$

then

$$r_{xx}(0) = \frac{1}{2\pi} \int_{-\infty}^{\infty} P_{xx}(\omega) \cdot d\omega$$

which is equal to $(1/2\pi)$ times the area under the signal's power spectral density characteristic. To summarise, if we know the spectral properties of a random input signal, we may estimate the changes to its mean and mean square value as it passes through a linear system having a known frequency response. This gives us information about the first and second moments of its amplitude distribution at the output— although only a complete set of moments would allow us to define the form of this distribution precisely.

There is one important type of amplitude distribution which is preserved by linear processing—the normal, or gaussian, distribution. In section 5.5.2, we noted that this form of distribution is specified by just two parameters, its mean value and variance. When a random signal with a normal amplitude distribution forms the input to a linear system, the output signal is also normal—although its mean and variance are generally different from those of the input (and so, of course, are its spectral properties). The reason for this important result may be understood by considering once again the graphical interpretation of the convolution integral. Suppose we have a normally-distributed sampled-data signal forming the input to a linear system with an impulse response $I(t)$: to evaluate the output at any instant t', we lay a reversed version of $I(t)$ beneath the input signal, cross-multiply, and sum all the terms—as illustrated by figure 7.24. Therefore each output is formed from a weighted set of input samples. But the sum of a number of normally-distributed variables is itself normal; hence the output signal will be normal in this case. Indeed, this arrangement suggests that, since the sum of a large number of random variables having any form of amplitude distribution tends to be normal (this was Gauss' original result), the amplitude distribution of a random output signal from a linear system will tend to be normal, even when the input is not.

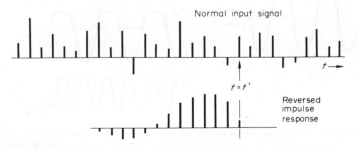

Figure 7.24

Although these arguments are simpler to appreciate in the case of random sampled-data signals, they apply equally well to continuous ones.

7.5   Nonlinear processing

In the foregoing discussion of signal processing, we have concentrated exclusively on the idea of linear systems. Such systems obey the principle of superposition, and have the important property of 'frequency-preservation': however different the input and output waveforms of a linear system may appear, the latter never contains frequencies which are absent from the input.

It is a sobering thought that, powerful though the concepts and techniques of linear signal processing may be, they exclude a great variety of signal operations of practical interest. Because there is no general theoretical framework available for the discussion of nonlinear processes, they are not covered in detail in this book. A nonlinear process does not obey the principle of superposition, nor does it have the property of frequency-preservation: nonlinear systems may not be completely characterised by impulse or frequency response functions, nor may their output signals be derived by transform methods or by convolution.

Certain relatively simple types of nonlinear process do not involve memory, implying that the instantaneous output depends only upon the instantaneous input[31]: amongst these processes are included half-wave and full-wave rectification, clipping, and squaring—all of which are shown in figure 7.25 applied to a sinusoidal input waveform. Processes involving memory, sometimes called 'reactive' processes, are more difficult to deal with analytically and include such functions as peak detection, in which the output signal is always equal to the largest positive (or negative) value so far attained by the input. Quite often, the output waveform

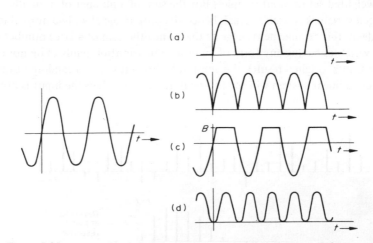

Figure 7.25   *Four examples of nonlinear processing. A sine wave (left) is*
(a) half-wave rectified, (b) full-wave rectified, (c) clipped at some level B, and
(d) squared

from a nonlinear process without memory may be most conveniently estimated by graphical methods; it is also sometimes useful to approximate a nonlinear input/output relationship by two or more linear characteristics, each of which applies to a certain range of input signal levels; and a reactive process may on occasions be separated into a nonlinear process without memory, plus a linear process with memory, which makes it rather easier to handle[31]. But the coherent body of knowledge which describes the effects of linear systems on signals in both time and frequency domains has no counterpart in the treatment of nonlinear signal processes.

Problems

1. A sinusoidal signal of frequency 1 kHz is applied as input to a signal processor with a frequency response

$$H(j\omega) = \frac{j\omega t_1}{1 + j\omega t_1}$$

where t_1 is a constant equal to 2×10^{-4} second. Estimate the modifications to the signal's amplitude and phase caused by the processor.

2. A square wave of period 1 second forms the input to a linear system with the frequency response illustrated in the accompanying figure. Estimate the relative magnitudes and phases of significant harmonic components at the output, and hence sketch one complete period of the output waveform.

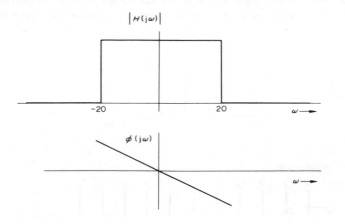

3. The transfer function of a linear system is given by

$$H(s) = \frac{1}{(s + 2)(s + 3)}$$

Plot its pole-zero configuration in the s-plane, and find the Laplace transform of its output signal when

(i) the input signal is a unit Dirac pulse applied at $t = 0$.
(ii) the input signal is the decaying exponential waveform e^{-4t}, applied at $t = 0$.

Using the table of Laplace transforms at the end of the text, find the output waveform in each case.

4. A linear system has the transfer function

$$H(s) = \frac{(s + 1)}{(s + 0.2 + 2j)(s + 0.2 - 2j)}$$

Write down an expression for its response to continuous sinusoidal inputs, and use it to estimate the magnitude and phase of the response to inputs of frequency $\omega = 0$, $\omega = 2$, and $\omega = \infty$. Why is its response magnitude so great when $\omega = 2$ radians/second?

5. Evaluate and sketch the impulse response of the system of problem 4. Why does the form of this response ensure that frequencies in the region of $\omega = 2$ radians/second will be strongly transmitted?

6. The impulse response and input signal of a sampled data system are illustrated in the accompanying figure. Estimate and sketch the form of the corresponding output signal. How long after the input has started does the output settle to a steady value? And how many nonzero output values are there after the input has ceased?

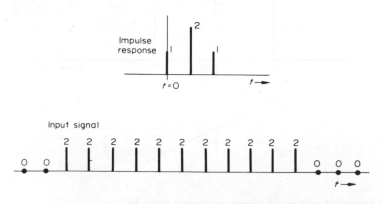

7. The impulse response and input signal of a continuous system are shown in the accompanying figure. Evaluate the output waveform using the convolution integral. What is the relationship between the form of the impulse response and that of the output waveform following the instant $t = t_1$?

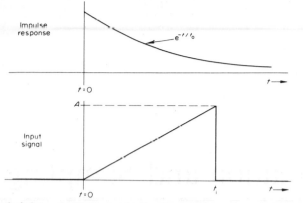

8. The sampled-data signal shown in the figure forms the input to a matched filter.

  (i) Sketch the filter's impulse response.
  (ii) Sketch the output signal, and satisfy yourself that it is identical in form to the autocorrelation function of the input.
  (iii) Write down an expression for the output energy spectral density, and sketch its form in the interval $0 < \omega < \pi/T$.

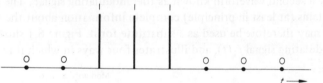

9. Wideband random noise having constant power spectral density P is present at the input of a linear system whose transfer function is

$$H(s) = \frac{1}{s^2 + 10s + 26}$$

Derive an expression for the noise power spectral density at the output, and sketch the result over a suitable range of frequencies.

10. The probability density function describing the amplitude of a random signal is shown in the accompanying figure. This signal forms the input to a linear system with a frequency response

$$H(j\omega) = \frac{1}{1 + j\omega t_1}$$

Estimate the average value, the mean square value, and the standard deviation of the random output signal.

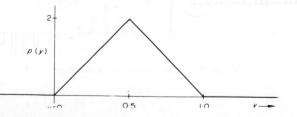

8

Modulation and Sampling

8.1 Introduction to modulation

In this chapter we investigate a number of signal techniques which fall under the general heading of 'modulation' processes.[5,7] The term 'modulation' implies the variation or adjustment of a property of one signal, generally referred to as the 'carrier', by a second waveform known as the 'modulating signal'. The modulated carrier contains (at least in principle) complete information about the modulating signal, and may therefore be used as a substitute for it. Figure 8.1 shows part of a typical modulating signal $f_1(t)$, and illustrates four ways in which it might be made

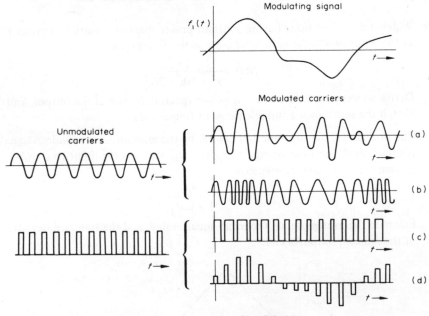

Figure 8.1 *Four types of modulation*

158

to modulate a carrier. In two cases, the carrier is assumed to be a relatively high frequency sinusoid, the amplitude or frequency of which is modulated in sympathy with the instantaneous value of the signal $f_1(t)$; and in the other two examples, $f_1(t)$ is made to modulate either the width or the height of successive pulses in a pulse-train.

There are various reasons why a modulated carrier might be used as a substitute for the modulating signal itself. Perhaps the best-known is the use of such a carrier for radio transmission; the carrier, being of high frequency, may be efficiently radiated from and received by suitable aerials, whereas the relatively low frequency modulating signal (which might, for example, represent music or speech) would require enormous aerials if it was to be transmitted directly. In magnetic tape recording, it is technically difficult to record very low frequency signals (say below a few hertz): but if such signals are used to modulate either the amplitude or frequency of a relatively high-frequency carrier, they may easily be recorded. Finally, a continuous signal may be converted into sampled-data form by causing it to modulate the amplitude of a train of narrow (Dirac) pulses: the sampled-data signal may then be stored or analysed by a digital computer, or transmitted as a set of numerical values rather than as a continuous waveform.

As the above discussion implies, one of the major effects of modulation is to shift the spectrum of the modulating signal—often from an essentially low-frequency region to a high-frequency one. The introduction of new frequencies is an essential property of any useful modulation process: such a process may never therefore be realised by a linear system with the property of frequency preservation, nor do modulation processes, as a general rule, fit into the framework of time and frequency descriptions of linear systems developed in the previous chapter. Fortunately, however, the type of modulation in which a signal causes variations in the instantaneous amplitude of a carrier is a notable exception, for reasons which will now be outlined.

The main point to note about 'amplitude modulation' is that the modulated carrier is formed by multiplying the modulating signal by the carrier. Hence if $f_c(t)$ represents the carrier prior to modulation, and $f_1(t)$ represents the modulating signal, the process of amplitude-modulation involves forming the product $f_c(t) \cdot f_1(t)$; this generates frequency components which are not to be found in either $f_c(t)$ or $f_1(t)$ alone. Modulated carriers (a) and (d) of figure 8.1 are examples of amplitude modulation. In section 7.2.3, we saw how multiplication of two functions of frequency is equivalent to convolution of their respective time functions. In fact, the convolution theorem is more general in scope than this; it merely states that the multiplication of two functions is equivalent to convolution of their Fourier transforms, regardless of what those functions represent. In amplitude modulation, it is the two time functions which are multiplied, and the convolution theorem therefore leads us to expect that this is equivalent to convoluting their individual frequency spectra. For these reasons, amplitude modulation is a rather special case, to which much of our earlier discussion of convolution and linear systems theory is relevant. We now look at this general concept in rather more detail, and apply it first of all to the process of signal sampling.

8.2   Signal sampling and reconstitution

8.2.1   The sampling process

Whenever a continuous signal is to be represented by a set of samples, a decision must be made about the sampling rate. If the sampling rate is too low, information about the detailed fluctuations of the continuous waveform will be lost: and if too high, an unnecessarily large number of samples will have to be stored or processed. We shall now show that the clue to an appropriate sampling rate lies in the relationship between the spectrum of a continuous signal and that of its sampled version.

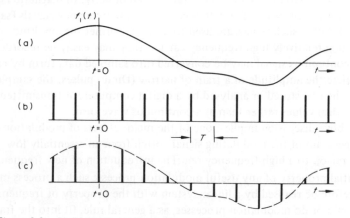

Figure 8.2   *Multiplication of a continuous signal (a) by a train of unit Dirac pulses (b) yields the sampled signal (c)*

The first point to make about the sampling process is that it is a form of amplitude modulation: from a mathematical point of view, sampling a continuous signal is equivalent to multiplying it by a train of equally-spaced unit Dirac pulses, as shown in figure 8.2. We may conveniently consider each sample to be a weighted Dirac pulse of standard width and a height (and hence area) proportional to the value of the continuous waveform at the relevant instant. Since the sampled signal is obtained by multiplication of the continuous waveform and the Dirac pulse train, its spectrum may be found by convoluting their respective spectra.

Our first task is therefore to define the spectrum of an infinite train of unit Dirac pulses, separated from one another by some sampling interval T. For convenience, let us assume that one of the pulses occurs at $t = 0$, as already shown in figure 8.2(b). Such a pulse train is a strictly periodic waveform, symmetrical about $t = 0$, and must therefore have a line spectrum containing only cosine terms: its fundamental frequency will clearly be $1/T$ hertz, or $2\pi/T$ radians/second. It remains to define the relative magnitudes of the various harmonics. Actually this is quite simply done, if we recall that the magnitude of any harmonic term in a periodic waveform may be found by multiplying the waveform by a cosine (or sine) of appropriate frequency, and integrating the product over one

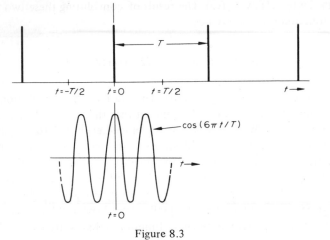

Figure 8.3

complete period. Figure 8.3 shows, as an example, part of a cosine of third harmonic frequency ($\omega = 6\pi/T$): if we multiply this by the pulse train waveform above, and integrate over the period $-T/2 < t < T/2$, the sifting property of the Dirac pulse (see section 4.2) ensures that the result will be simply equal to the value of the cosine wave at $t = 0$, which is unity. Indeed this same result applies to any other harmonic frequency (and for the zero-frequency term) because cosines of all frequencies have a value of unity at $t = 0$. We conclude that the spectrum of the pulse train of figure 8.3(a) contains an infinite set of cosine harmonics, all of the same amplitude and separated by $2\pi/T$ radians/second, as shown in figure 8.4. This result emphasises a special property of the Dirac pulse train—that its time-domain waveform and its frequency spectrum are identical in form. It is perhaps unsurprising that its spectrum extends to infinitely high frequencies, because it is made up from Dirac pulses which are themselves completely 'wideband'; and since the time-function is strictly repetitive, the spectrum can contain only discrete harmonic frequencies.

The next problem is to investigate what happens when the spectrum of figure 8.4 is convoluted with that of a typical continuous signal, for such a convolution will yield the spectrum of the signal's sampled version. Let us denote the spectrum of the continuous signal by $G_1(\omega)$ and the spectrum of the pulse train (already

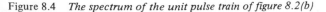

Figure 8.4   *The spectrum of the unit pulse train of figure 8.2(b)*

illustrated in figure 8.4) by $G_2(\omega)$. The result of convoluting these two functions is a further function $G_3(\omega)$, given by

$$G_3(\omega) = \int_{-\infty}^{\infty} G_1(\omega - \Omega) . \, G_2(\Omega) . \, d\Omega$$

where Ω is an auxiliary frequency variable. This form of the convolution integral is identical to that used in section 7.2.3, except that time functions are replaced by frequency functions. We should of course remember that $G_1(\omega)$ is generally complex, whereas $G_2(\omega)$ is purely real since it is the spectrum of an even time function. However, the above convolution integral is simpler to visualise if we think of both G_1 and G_2 as real functions of ω; and the general validity of the result is not affected by this assumption. Now as we have already shown, the spectrum $G_2(\omega)$ consists of a series of equally-spaced spectral lines, which may also conveniently be represented by Dirac functions (that is, frequency-domain 'pulses'). Hence

$$G_2(\omega) = \ldots + \delta\left(\omega + \frac{4\pi}{T}\right) + \delta\left(\omega + \frac{2\pi}{T}\right) + \delta(\omega) + \delta\left(\omega - \frac{2\pi}{T}\right) + \delta\left(\omega - \frac{4\pi}{T}\right) + \ldots$$

and therefore

$$G_3(\omega) = \int_{-\infty}^{\infty} G_1(\omega - \Omega) \left\{ \ldots \delta\left(\Omega + \frac{4\pi}{T}\right) + \delta\left(\Omega + \frac{2\pi}{T}\right) + \delta(\Omega) \right.$$

$$\left. + \delta\left(\Omega - \frac{2\pi}{T}\right) + \delta\left(\Omega - \frac{4\pi}{T}\right) + \ldots \right\} . \, d\Omega$$

$$= \ldots + \int_{-\infty}^{\infty} G_1(\omega - \Omega) . \delta\left(\Omega + \frac{4\pi}{T}\right) . \, d\Omega + \int_{-\infty}^{\infty} G_1(\omega - \Omega) . \delta\left(\Omega + \frac{2\pi}{T}\right) . \, d\Omega$$

$$+ \int_{-\infty}^{\infty} G_1(\omega - \Omega) . \delta(\Omega) . \, d\Omega + \ldots$$

Fortunately, the sifting property of the Dirac function makes the various terms of this integral very simple to find. For example, the function $\delta(\Omega + 4\pi/T)$ represents a Dirac 'pulse' at $\Omega = -(4\pi/T)$, and when this is multiplied by $G_1(\omega - \Omega)$ and integrated, the result is just the value of $G_1(\omega - \Omega)$ at $\Omega = -(4\pi/T)$, namely $G_1(\omega + 4\pi/T)$. Hence

$$G_3(\omega) = \ldots G_1\left(\omega + \frac{4\pi}{T}\right) + G_1\left(\omega + \frac{2\pi}{T}\right) + G_1(\omega)$$

$$+ G_1\left(\omega - \frac{2\pi}{T}\right) + G_1\left(\omega - \frac{4\pi}{T}\right) + \ldots$$

Now a function $G_1(\omega + 4\pi/T)$ is identical to $G_1(\omega)$ apart from a bodily shift of $4\pi/T$ radians/second along the frequency axis: therefore the result of convoluting $G_1(\omega)$ and $G_2(\omega)$ is a function $G_3(\omega)$ in which $G_1(\omega)$ is repeated indefinitely along the frequency axis at intervals of $2\pi/T$ radians/second. (This illustrates the general fact that whenever a function is convoluted with a Dirac pulse train the effect is to repeat the function at intervals equal to the pulse spacing.) Figure 8.5 shows a typical continuous signal spectrum $G_1(\omega)$, and that of its sampled version $G_3(\omega)$, when the sample spacing is T seconds. Strictly speaking, the peak values of $G_3(\omega)$

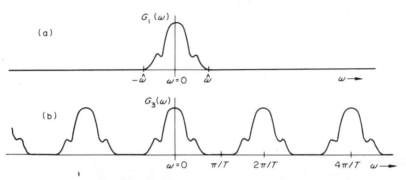

Figure 8.5   (a) The spectrum of a typical continuous signal and (b) that of its sampled version when the sampling interval is T seconds

should be shown much smaller than that of $G_1(\omega)$; as the figure stands, it falsely suggests that the total energy in $G_3(\omega)$ is greater than that in $G_1(\omega)$.

This important result shows that the spectrum of a sampled signal is a repeated version of that of the underlying continuous waveform: it also gives a vital clue to the minimum sampling rate which may be used if the sample values are to form an adequate substitute for the original signal. Figure 8.5 has shown a spectrum $G_1(\omega)$ with significant frequency components in the frequency range $-\hat{\omega}$ to $+\hat{\omega}$: sampling causes this spectrum to repeat every $2\pi/T$ radians/second. It is therefore clear that if $\hat{\omega}$ is greater than π/T there will be overlap between adjacent repetitions of $G_1(\omega)$. Without such overlap it would be possible, at least in principle, to recover the original continuous waveform from its sampled version by passing the latter through a linear filter which transmitted equally all components in the range $-\hat{\omega}$ to $+\hat{\omega}$, and rejected all others. But with such overlap, the spectrum of the sampled signal is no longer simply related to that of the original (especially in the regions of $\pm\hat{\omega}$), and no linear filtering operation could be expected to recover the original from its sampled version. The overlap which arises when the sampling rate is too low is often referred to as 'aliasing'.

The above argument specifies, in effect, a minimum sampling rate for the faithful representation of a continuous signal by a sample set. By 'faithful representation' we imply that the sample values contain complete information about fluctuations in the original signal, and that they could therefore be used to reconstitute the original if required. It is at first sight remarkable that a set of samples can ever form a complete substitute for a continuous waveform, for it seems inevitable that some details of the latter's fluctuations must be lost in the sampling process. The reason why this is not so is that we have assumed an upper limit ($\hat{\omega}$) to the frequencies contained in the continuous waveform, which is equivalent to specifying a maximum rate at which it can change its value. If its maximum rate of change is limited, then all possible fluctuations can be detected by taking samples at suitably-spaced instants.

The minimum sampling rate for adequate representation of a continuous signal is formally specified in the so-called 'sampling theorem'. If $\hat{f}$ is the frequency in hertz corresponding to $\hat{\omega}$ radians/second, then $\hat{f} = \hat{\omega}/2\pi$: and if the 'overlap' condition is to be avoided we have already shown that

$$\hat{\omega} < \frac{\pi}{T}$$

or

$$\hat{f} = \frac{\hat{\omega}}{2\pi} < \frac{1}{2T}$$

and hence

$$T < \frac{\pi}{\hat{\omega}} = \frac{1}{2\hat{f}}$$

Formally, the sampling theorem may therefore be stated as follows:

'A continuous signal which contains no significant frequency components above $\hat{f}$ hertz may in principle be recovered from its sampled version, if the sampling interval is less than $1/2\hat{f}$ seconds'.

The above discussion of the relationships between a continuous signal and its sampled version throw considerable light on some of the points first mentioned in chapter 4. There we noted that the frequency spectrum of a sampled data signal is always repetitive in form, and we now recognise this to be an essential characteristic of the sampling process. It is also now clear why a certain minimum sampling rate must be used if a sampled-data signal is to represent a continuous underlying waveform adequately; and why, when adequate sampling has been used, that portion of the sampled-data signal's spectrum lying in the range $-\pi/T < \omega < \pi/T$ may be used to define both the sampled-data signal itself, and the underlying waveform which it represents.

8.2.2 Signal reconstitution

We next examine in rather more detail the idea of reconstituting the original continuous signal from its sample values. The process of reconstitution is illustrated in figure 8.6. In part (a) the spectrum of a sampled signal is shown: the maximum frequency in the original signal is $\hat{\omega}$, and 'overlap' is just avoided by use of a sampling interval of $(\pi/\hat{\omega})$ seconds. Also shown is the frequency-response

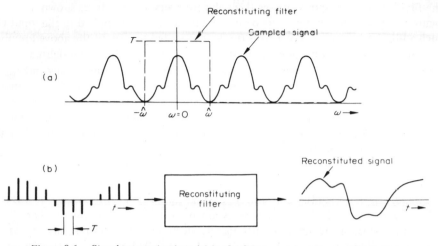

Figure 8.6 *Signal reconstitution, (a) in the frequency-domain, and (b) in the time-domain*

characteristic of the required reconstituting filter. Part (b) of the figure shows the equivalent time-domain operation: the sampled version of the signal forms the input to this filter, whose output is the original continuous waveform.

We may obtain more insight into the reconstitution process if we consider the impulse response $I(t)$ of the filter. This is given by the transform of its frequency response, hence

$$I(t) = \frac{1}{2\pi} \int_{-\infty}^{\infty} G(j\omega) \cdot e^{j\omega t} \cdot d\omega = \frac{1}{2\pi} \int_{-\hat{\omega}}^{\hat{\omega}} T \cdot e^{j\omega t} \cdot d\omega$$

$$= \frac{T}{2\pi} \cdot \frac{1}{jt} (e^{j\hat{\omega}t} - e^{-j\hat{\omega}t})$$

$$= \frac{T}{2\pi} \cdot \frac{1}{jt} \cdot 2j \sin \hat{\omega}t = \frac{T \sin \hat{\omega}t}{\pi t} = \frac{\sin \hat{\omega}t}{\hat{\omega}t}$$

$I(t)$ is of the familiar $(\sin x/x)$ form, and passes through zero whenever $\hat{\omega}t = n\pi$, or $t = n\pi/\hat{\omega}$, where n is an integer. But $(\pi/\hat{\omega})$ is the sampling interval T: therefore

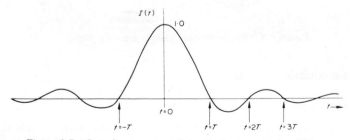

Figure 8.7 *Impulse response of the ideal reconstituting filter*

the filter's impulse response passes through zero every T seconds, as shown in figure 8.7. When the sampled version of the original signal is applied to the input of such a filter, each sample value gives rise to a weighted version of the above impulse response. The filter output is equal to the superposition of all these weighted impulse responses. It is interesting to note that the filter's response at any one sampling instant is not affected by adjacent samples, because the impulse response passes through zero every T seconds. This is surely what we should expect: after all, each sample value is a true value of the original continuous waveform and the reconstituted waveform must pass exactly through it; it will not do so unless impulse response contributions from adjacent samples are zero.

Unfortunately, the ideal reconstituting filter does not represent a physically realisable device, because its impulse response begins before the instant at which the impulse is delivered (by convention, $t = 0$ in figure 8.7). The reasons for this difficulty are twofold: firstly, we have assumed a filter with an infinitely sharp 'cut off', that is, one which transmits equally all frequencies up to $\hat{\omega}$, and rejects completely all higher ones; and secondly, we have implied no phase shift in those components transmitted, because $G(j\omega)$ has been put equal to T, a real constant. As we shall see in the next chapter, no realisable filter can achieve an infinitely sharp cut-off, nor can it fail to introduce phase shift. The best we can hope to achieve in practice is a reasonably sharp cut-off beginning at $\hat{\omega}$, and a phase shift roughly proportional to frequency—which implies a more or less constant delay imposed upon all components passing through the filter. Our reconstituted signal will then be a fairly faithful, but somewhat delayed, version of the original signal prior to sampling. The similarity of the reconstituted and original signals may be enhanced by using a sampling rate somewhat higher than that specified by the sampling theorem. This has the effect of introducing a 'guardband' in the spectrum of the sampled signal and the reconstituting filter may then be designed so that its response falls to a very small value as the guardband is crossed. These points are illustrated by figure 8.8.

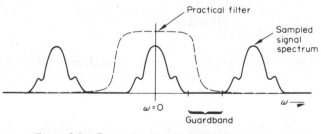

Figure 8.8   *Reconstitution using a practical filter*

8.3   Other modulation processes

8.3.1 Signal truncation

Suppose we have a periodic signal $f(t)$ with a line spectrum $G(j\omega)$, illustrated in figure 8.9(a). If the period of $f(t)$ is T_0 seconds, successive harmonic terms are spaced $(2\pi/T_0)$ radians/second apart. (For the purpose of our discussion we assume $f(t)$ to be an even function, and therefore $G(j\omega)$ to represent simply the magnitudes of the various cosine harmonics.) Next suppose that we 'truncate' $f(t)$, by assuming it to be zero outside the limits $t = \pm T_0/2$. Physically, such truncation might represent the fact that we only observe or record $f(t)$ during this limited interval, and have

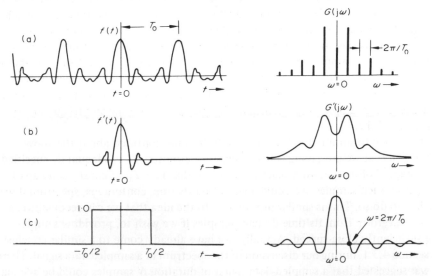

Figure 8.9   *Signal truncation in the time and frequency domains*

no knowledge of its value at other times. Let us call the truncated version $f'(t)$, and its spectrum $G'(j\omega)$: the latter will now, of course, be continuous, since $f'(t)$ is an aperiodic function. $f'(t)$ and $G'(j\omega)$ are shown in figure 8.9(b).

Because $f(t)$ and its truncated version are closely related, so must be their corresponding spectra, and it is of considerable interest to define the precise nature of this relationship. Mathematically, the truncation process is equivalent to multiplying the original repetitive signal by the rectangular pulse of figure 8.9(c), and it may therefore be thought of as a type of amplitude modulation (even if we cannot sensibly describe one of the waveforms as the 'carrier', and the other as the 'modulating signal'). For reasons mentioned above, such a rectangular pulse is sometimes referred to as an 'observation window'. Now multiplication of two time functions is equivalent to convolution of their spectra: therefore the spectrum of $f'(t)$ may be found by convoluting the line spectrum of $f(t)$ with the spectrum of the rectangular pulse function. This latter spectrum is of the familiar $(\sin x/x)$ form, and is also shown in part (c) of figure 8.9. When a line spectrum is convoluted with a spectrum of $\sin x/x$ form, the result is to repeat a weighted version of the $\sin x/x$ pattern about each of the lines. In other words the spectrum of the truncated signal $f'(t)$ may be found by arranging a $\sin x/x$ function around each of the lines in the spectrum of $f(t)$ itself. We have now defined exactly the relationship between these two spectra.

This result implies that if we know the line spectrum of the original repetitive signal, we can find that of its truncated version; or conversely, that the spectrum of the truncated signal is adequately represented by the line spectrum. This conclusion is in a striking sense analogous to that of the sampling theorem. The sampling theorem states that a signal whose spectral energy is limited to the range $-\hat{\omega} < \omega < \hat{\omega}$ may be adequately represented by a set of sample values: we now find that a signal which is time-limited (that is, truncated so that it is zero outside specified time-limits) may be adequately represented by a set of spectral lines— even though its spectrum is in fact continuous. We may therefore think of the spectral lines as 'samples' of the underlying continuous spectrum. Perhaps it should come as no surprise to find that the sampling theorem, normally applied to the time-domain sampling of a waveform, has its counterpart in the frequency-domain: it is a further example of the constant interplay between time and frequency in linear signal theory.

Let us try to summarise, for it is easy to become confused about the above result. All that we are saying is that a signal of duration T_0 is adequately specified by 'samples' of its spectrum spaced (not more than) $2\pi/T_0$ radians/second apart: and, given such samples, we could always find its true, continuous, spectrum if we wished to do so. This is similar in principle to the idea that we can reconstitute a signal waveform from its time-domain samples if we wish to, providing samples are sufficiently close-spaced. Actually, we have already come to a similar conclusion in section 4.3.1, during our discussion of the spectrum of a sampled data signal. There, it was suggested that a sampled-data signal of duration N samples could be adequately described by a spectrum containing $N/2$ harmonics. It is not difficult to show that this is equivalent to specifying a signal of duration T_0 seconds by a set of spectral lines spaced $2\pi/T_0$ radians/second apart.

Truncation of a signal always tends to cause spreading of its spectral energy. This is most clearly seen in the type of case we have investigated, in which truncation causes a line spectrum to change to a continuous one. The more severe the truncation (that is, the narrower the 'observation window'), the more pronounced is the spreading effect. Sometimes the effect is undesirable; at other times it may be useful. For example, if we wish to evaluate the spectrum of a signal and to distinguish the relative magnitudes of closely spaced frequency components, the spreading or smearing effect of truncation will foil us if we process too short a signal portion. On the other hand, it may be helpful to have the short-term fluctuations in a spectrum smoothed out, particularly if they represent statistical fluctuations to which we can attach little significance. An example of this arises in the practical estimation of the power spectrum of a random signal (a problem already referred to in section 5.4.4). Rather than estimate the power spectrum of a limited portion of a random signal and then smooth the resulting curve, it is generally simpler and faster to estimate a truncated version of its equivalent time function—the autocorrelation function. The latter then transforms directly into the required smoothed version of the power spectrum. Actually, the question of how best to truncate a time-domain function in order to achieve satisfactory smoothing of its corresponding spectrum is one of considerable practical interest, and is covered in more advanced texts on the analysis of random signals. Suffice it to note here that the rectangular 'observation window' is not generally the most satisfactory from this point of view, and that time functions are quite often multiplied by other forms of window (such as the well-known Hanning function) before transformation into the frequency-domain.[32]

8.3.2 Amplitude modulation

We have already discussed two types of modulation process—signal sampling and truncation—which may be thought of as examples of amplitude modulation. We now turn our attention to the conventional type of amplitude modulation[7] (AM) which employs a high-frequency sinusoidal waveform as the carrier. This type of modulation has already been illustrated in part (a) of figure 8.1.

To simplify the discussion, suppose the modulating signal is a simple cosine wave $\cos \omega_m t$ and the carrier is given by $\cos \omega_c t$: in most practical cases the ratio (ω_c/ω_m) is at least 20 so that one period of the modulating wave encompasses many periods of the carrier. Modulation of the carrier involves forming the product

$$\cos \omega_m t \,.\, \cos \omega_c t$$

which, by using the well-known trigonometrical identity, may be written as

$$\tfrac{1}{2} \cos (\omega_c + \omega_m) t + \tfrac{1}{2} \cos (\omega_c - \omega_m) t$$

This result shows that the modulated carrier contains neither the frequency ω_c nor the frequency ω_m, but only the sum and difference frequencies $(\omega_c + \omega_m)$ and $(\omega_c - \omega_m)$. It is a further example of the introduction of new frequencies by a modulation process. In practice, the modulating signal will normally contain a range of frequencies, each of which will give rise to its own sum and difference frequencies in the spectrum of the modulated carrier. The resulting 'sidebands', arranged on either side of the carrier frequency ω_c, are shown in figure 8.10. The spectrum of the modulating signal has been translated from the region around $\omega = 0$ to the region around $\omega = \pm \omega_c$.

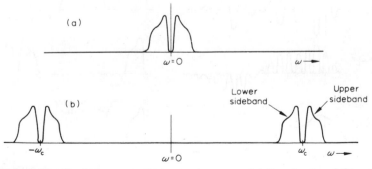

Figure 8.10   *Typical spectra of (a) a continuous modulating signal, and (b) an amplitude-modulated carrier*

No direct reference has so far been made to the process of 'demodulation', which consists of retrieving the modulating signal from the modulated carrier. Just as amplitude modulation involves shifting the signal to a relatively high-frequency region, so demodulation involves shifting it back again. In a radio transmission system, demodulation of the modulated carrier is, together with amplification, the major task of the receiver. We have just seen how multiplication of the modulating signal and carrier gives rise to sum and difference frequencies: not surprisingly, a further multiplication of the modulated carrier by a wave at carrier frequency causes the generation of new sum and difference frequencies, amongst which are included those of the original signal. Considering typical components $(\pm\omega_c \pm \omega_m)$ present in the modulated carrier, multiplication by a cosine wave at carrier frequency ω_c produces the new sum and difference frequencies

$$(\pm 2\omega_c \pm \omega_m)$$

and

$$(\pm\omega_m)$$

of which those at $\pm\omega_m$ represent the original modulating signal. The other components, which are arranged around the second harmonic of the carrier frequency, may quite easily be removed by a suitable filter. The complete sequence of modulation and demodulation is illustrated in figure 8.11.

In practice, demodulation is not quite such a straightforward operation as the above argument suggests. The reason for this is that it involves multiplying the

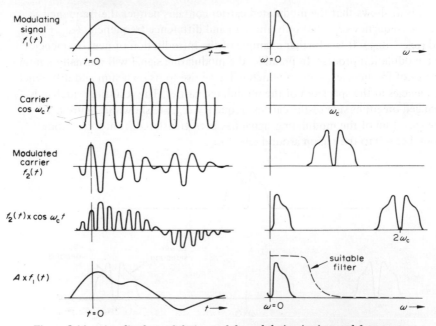

Figure 8.11 *Amplitude modulation and demodulation in time and frequency domains. The bottom two diagrams represent the demodulation process*

received modulated carrier by the function cos $\omega_c t$; since the modulated carrier contains no component at frequency ω_c, this means that a cosine wave at carrier frequency must be generated in the receiver, and this is not a simple matter. In order to simplify the demodulation process, most broadcasting systems transmit a wave at carrier frequency in addition to the sidebands containing the signal information. Since the transmission of this carrier frequency component is quite normal practice, the process is often referred to as 'normal' amplitude modulation; when the carrier component is not transmitted, the process is known as 'suppressed carrier' amplitude modulation.

8.3.3 Frequency and pulse-code modulation

In frequency modulation, the instantaneous frequency of the carrier wave is varied in sympathy with the modulating signal; this situation has already been illustrated in figure 8.1(b). Frequency modulation (FM) is used for high-quality broadcasting and, occasionally, for the faithful recording of low-frequency signals on magnetic tape. Its main advantage over the more conventional AM techniques lies in its potential freedom from noise and interference[7].

Another important class of modulation processes involves the use of a pulse train: the width of pulses can be varied in sympathy with the modulating signal ('pulse-width modulation', shown in figure 8.1(c)); or their height may be varied (as in signal sampling). An interesting variation of this latter scheme is pulse-code modulation (PCM)[33], which is of increasing practical importance for the transmission of signals by radio or cable. In PCM, the continuous signal is first represented by a set of sample values, which are then transmitted in the form of a pulse code. Such a scheme is illustrated in figure 8.12. Each sample value is represented in the transmitted code by a sequence of, say, five binary pulses: such a code is able to

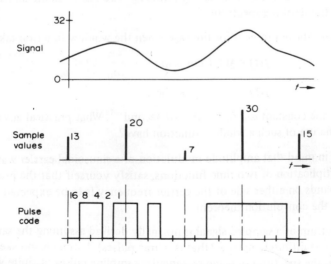

Figure 8.12 *Pulse-code modulation (PCM). The five binary pulses which occupy each sampling interval are used to denote the preceding sample value*

handle only $2^5 = 32$ distinct levels, which means that each measured sample value has to be rounded off to the nearest level before coding takes place. The more binary pulses are allocated to each sample value, the more accurate is its representation by the pulse code. The process of rounding-off the sample values is referred to as 'quantisation', and it causes inevitable errors in the coded version of the signal. In spite of this drawback, the use of PCM is increasing for two main reasons: firstly, recent advances in electronics make it technically more attractive than it was when first investigated in the 1930s; and secondly, it offers great resistance to further signal degradation during transmission—mainly because the receiver has to distinguish only two possible (binary) pulse levels, rather than the many signal levels of the conventional amplitude modulated carrier.

These brief notes on frequency and pulse-code modulation have been included to avoid giving the impression that amplitude modulation is the only major type of modulation process in common use. Such an impression would be quite erroneous, although it is true that amplitude modulation—in its various guises—is very important. What is perhaps more significant in the present context is that amplitude modulation fits broadly into the framework of linear signal analysis and processing developed in earlier chapters of this book, because it involves multiplication of the modulating signal and the carrier. This is definitely not the case with other processes such as frequency, pulse width, or pulse-code modulation, and the time and frequency-domain descriptions of these latter processes are therefore much more difficult to relate.

Problems

1. A continuous cosinusoidal waveform of frequency 100 Hz is truncated by a rectangular window function having limits $t = \pm 50$ ms. Estimate and sketch the form of the resulting spectrum.

2. Repeat the above problem for the case when the window function takes the form

$$f(t) = \tfrac{1}{2}(1 + \cos \omega_0 t), \qquad \frac{-\pi}{\omega_0} < t < \frac{\pi}{\omega_0}$$

$$f(t) = 0, \qquad \text{elsewhere.}$$

Assume the constant ω_0 to equal 20π second^{-1}. What practical advantages might the use of such a window function have?

3. Bearing in mind that amplitude modulation of a sinusoidal carrier wave involves the multiplication of two time functions, satisfy yourself that the production of sidebands on either side of the carrier frequency is to be expected on the basis of the convolution theorem.

4. The spectrum of a sampled signal is normally derived assuming the sampling pulses to be infinitely narrow (that is, Dirac pulses). Estimate and sketch the effect on the spectrum of using rectangular sampling pulses of finite width t_0 seconds. What is the practical significance of your result?

9

Filters

9.1 Introduction

Linear filters form a class of system which is of crucial importance in signal processing. Although in its most general sense the term 'filter' implies any frequency-selective device or processor, in practice it is generally reserved for a system which transmits a certain range (or ranges) of frequencies, and rejects others: such frequency ranges are called 'passbands' and 'stopbands' respectively. We shall see later that the ideal filter, which would introduce no attenuation of input signals falling within the passband, and infinite attenuation of signals in the stopband, is not attainable in practice.

Historically, both the theory and practical application of filters have been very much tied up with electronic communications. For example, a radio receiver is required to discriminate in favour of just one of the many incoming signals picked up by its aerial; this it does on the basis of their different frequency bands, by use of a highly selective filter. Such a filter processes continuous signals and is therefore an example of what we have previously called a 'continuous' linear system; it is also widely referred to as an 'analogue' filter. Analogue filters are invariably constructed from linear electrical circuit components, and their detailed design falls outside the scope of this book; on the other hand we are in a position to discuss the overall performance of certain well-known types of analogue filter, and this is done in section 9.3.

Although a knowledge of electrical network theory is needed for the design of analogue filters, the same is fortunately not true of filters for sampled-data signals. Sampled-data filters, generally known as 'digital' filters, may be realised by suitable programming of a digital computer which is fed with a sampled version of the input signal. The increasing interest in digital filters is largely a reflection of the availability of the digital computer as a research tool in all branches of science and technology. The work we have done on the z-transform in chapter 4 and on linear systems in chapter 7 forms an adequate background for the design and implementation of digital filters. However before getting involved in detail, we

173

first investigate some general aspects of filter performance in time and frequency domains.

9.2   General aspects of filter performance

9.2.1   Filter categories

Apart from the division of linear filters into the two broad categories of analogue and digital filters, they may be further classified according to the frequency ranges which they transmit or reject. A 'low-pass' filter has a passband in the low-frequency region, whereas a 'high-pass' filter transmits only high-frequency input signals; 'band-pass' and 'band-stop' filters are defined by their ability to discriminate in favour of, or against, particular frequency bands. The actual frequency (or frequencies) at which the transition from passband to stopband occurs varies from case to case, and is clearly an important parameter of filter design. Idealised forms of these various classes of filter are illustrated in figure 9.1.

Since the frequency response of a linear sampled-data system is always a periodic function of ω (repeating itself every $2\pi/T$ radians/second, where T is the sampling interval), it follows that the terms low-pass, high-pass, bandpass and bandstop have to be interpreted slightly differently in the case of digital filters. We have already noted in section 8.2.1 that sampling with an interval of T seconds allows faithful representation of a continuous signal having frequency components up to $\hat{\omega} = \pi/T$ radians/second. A digital filter is therefore classified according to its effect on frequency components in the range $-\pi/T < \omega < \pi/T$, which is the maximum range occupied by any adequately-sampled input signal. Typical idealised digital filter responses are shown in figure 9.2.

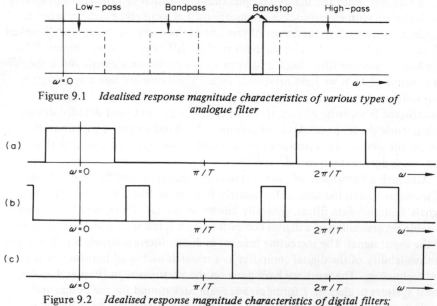

Figure 9.1   *Idealised response magnitude characteristics of various types of analogue filter*

Figure 9.2   *Idealised response magnitude characteristics of digital filters; (a) low-pass, (b) bandpass, and (c) high-pass*

9.2.2 Response in time and frequency domains

Like any other linear system, a frequency-selective filter may be described either
by its frequency response (or transfer function), or by its impulse response. The
frequency response description is the one normally used, because a filter is generally
specified in terms of its ability to discriminate against certain frequency ranges
and in favour of others; but any form of frequency response implies a particular
shape of impulse response, and the latter gives some important clues to filter
performance.

As a rule, the more limited the band of frequencies transmitted by a filter, the
more extended in time is its impulse response waveform: this is just a reflection of
the general antithesis between frequency-limitation and time-limitation, discussed
with reference to signal waveforms in section 3.3.2. It means that the output from
a highly selective filter must always be expected to take a long time to settle to
zero after the input has been removed, and that its response to a sinusoidal input
or steady level will take a long time to reach its 'steady state' after the input has
been applied. The transient effects which accompany the sudden application of an
input signal to a selective filter are sometimes referred to as 'ringing'. Since the
frequency response and impulse response of a filter are related as a Fourier transform pair, the form of the impulse response must be expected to reflect those
frequencies which are strongly transmitted by the filter. For example the impulse
response of figure 9.3(a) has the form of a decaying oscillation at a frequency of
about ω_0 radians/second, with zero average value: the corresponding frequency
response will therefore display strong transmission in the region of $\omega = \omega_0$, and
rejection of zero-frequency inputs. Another filter which displays a more prolonged
oscillatory impulse response has the more selective bandpass characteristic shown
in figure 9.3(b).

The reason why no linear filter can display the ideal characteristics shown in
figures 9.1 and 9.2 (a fact which is formally expressed by the co-called Paley-Wiener
criterion[1]) becomes clear if it is recalled that any linear system has a transform
function expressible in terms of a set of poles and zeros. As we saw in section 7.2.2,

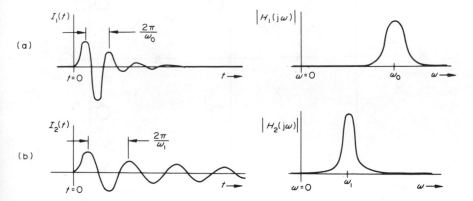

Figure 9.3   *Impulse and frequency responses of two analogue bandpass filters*

given a set of s-plane poles and zeros it is possible to infer the system's response to any sinusoidal frequency by drawing vectors from the various poles and zeros to a point on the imaginary axis in the s-plane (or to a point on the unit circle in the z-plane in the case of a sampled-data system). The response is large if the product of 'zero-vector' magnitudes is large, and/or the product of 'pole-vector' magnitudes is small. An infinitely fast transition from passband to stopband therefore implies an infinite rate of change of one or both of these products as a particular point on the imaginary axis is crossed. It is intuitively clear that such an effect cannot be achieved by any arrangement of a finite set of poles and zeros—although it may be more closely approximated when a large number of poles and zeros is specified. However, the number of poles and zeros in a transfer function reflects the complexity of the system, and complexity is closely related to cost.

The pole-zero approach may also be used to demonstrate the relationship between the magnitude and phase responses of a linear filter.[17] Let us start by considering an analogue filter which has all its poles and zeros in the left-hand half of the s-plane, as in figure 9.4(a). This configuration gives rise to a particular magnitude and phase response, and the two will be interdependent. It is interesting to consider whether the phase response may now be adjusted independently of the magnitude response. Actually this may be done in two ways, illustrated in parts (b) and (c) of the figure. In part (b) the zeros of the transfer function are moved across the imaginary axis to mirror-image positions in the right-half s-plane. The length of the zero vectors drawn to any point on the imaginary axis is clearly unaltered by this move, but their contribution to the phase response is changed. In part (c) of the figure, the original pole-zero pattern is augmented by additional pairs of poles and zeros arranged symmetrically with respect to the imaginary

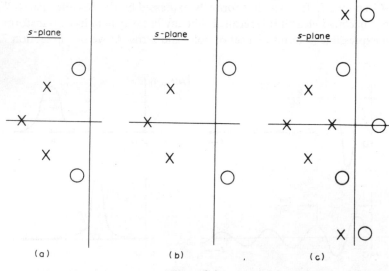

(a)   (b)   (c)

Figure 9.4

axis. These additional pairs cause no alteration to the magnitude response of the filter, since variations in length of one of the new zero vectors are exactly counter-balanced by those of the corresponding pole vector; on the other hand, their phase contributions do not cancel (except at very high frequencies). We should note that phase variation cannot be obtained by placing poles in the right-half s-plane, because this would give rise to an unstable filter. From this brief discussion, we may draw several conclusions. Firstly, if zeros are to be restricted to the left-half s-plane, magnitude and phase response are uniquely related, and may not be adjusted independently: a filter of this type is known as a 'minimum-phase' system. If a filter transfer function has zeros (and many functions of practical interest do not, as we shall see later), these may be placed in either the left or right-half s-plane, giving a corresponding flexibility in the phase response, but without altering the magnitude response. And finally, the phase response may be adjusted by adding mirror-image pole-zero pairs, although this adds to the filter's complexity and should therefore be avoided if possible. In practice, adjustment of the phase response of a filter is often accomplished by a separate system known as an 'all-pass network', which provides, in effect, just the required mirror-image pole-zero pairs. Such a process is sometimes referred to as phase 'equalisation'. However, it is fortunately true that the phase response of a filter is relatively unimportant in many applications; filter design therefore tends to concentrate on magnitude response, and the phase response is left to look after itself.

Although the above discussion of pole-zero configurations has been limited to analogue filters and the s-plane, the same general concepts apply equally well to digital filters and the z-plane, provided we remember that the imaginary axis in the s-plane is equivalent to the unit circle in the z-plane. Furthermore, since $z = e^{sT}$ (where T is the sampling interval of the system) a 'mirror image' pole-zero pair placed at $s = (\pm\sigma + j\omega)$ has its z-plane counterpart at

$$z = e^{(-\sigma + j\omega)T} = e^{-\sigma T}.e^{j\omega T}$$

and

$$z = e^{(\sigma + j\omega)T} = e^{\sigma T}.e^{j\omega T}$$

Thus the zero lies at a distance $e^{-\sigma T}$, and the pole at a distance $(e^{\sigma T})$, from the origin: both also lie on a radius vector which makes an angle ωT radians with the positive real axis. These equivalent s-plane and z-plane locations have already been illustrated in figure 4.7.

9.3 Analogue filters

9.3.1 General

In principle, it is possible to realise any desired s-plane pole-zero configuration using an electrical network. However the number of poles and zeros used bears a direct relationship to filter complexity, and hence cost, so that it is desirable to achieve an acceptable filter performance using as few poles and zeros as possible. In this section on analogue filters, we discuss briefly the pole-zero configuration and frequency responses of some common types of filter, each of which represents a useful compromise between ideal performance and design economy. Our discussion will be somewhat biased towards filters having low-pass characteristics; however, it is generally possible to convert a low-pass filter into a high-pass, bandpass, or bandstop one with similar passband and stopband performance, by modifications to the electrical circuit components. The mathematical techniques involved in such conversions fall under the heading of 'frequency transformations'[34], and are part of the stock-in-trade of the analogue filter designer.

What might be termed 'conventional' analogue filters are composed of passive linear electrical elements—resistors, inductors, and capacitors. In the design of such filters account must generally be taken of the electrical impedance of the devices to which the filter is to be connected; in other words the performance of the filter is generally affected by the electrical characteristics of the signal source connected to its input side, and of the 'load' (which might be another processor or some sort of recording device) connected to its output. Due attention must be paid to this question, because incorrect termination of an analogue filter may lead to serious discrepancies between its advertised and actual frequency responses. In more recent years, 'active' filters[35] have become increasingly popular. An active filter incorporates one or more active devices— such as amplifiers—which need a power supply, and its main advantages over the more conventional passive filter are twofold: the use of active devices allows inductors to be eliminated from the network, and inductors tend to be bulky, expensive, and difficult to realise in ideal form, particularly when the filter is to work at low frequencies; furthermore, it is a relatively simple matter to design an active filter which is insensitive to the precise electrical characteristics of the signal source and load to which it is connected.

9.3.2 Some common filter types

Filters composed entirely of inductors and capacitors, sometimes referred to as 'reactive' filters[8], have transfer function poles and zeros which are restricted to the imaginary axis in the s-plane. The theory of such filters has been extensively investigated ever since the early days of radio, and they have found widespread practical application. The pole–zero configuration and frequency response characteristic of a typical bandpass filter of this type are shown in figure 9.5. These are theoretical diagrams based upon ideal filter components and ideal terminations, neither of which may be realised in practice. The effect of practical components and terminations (which always possess electrical resistance) is to move the poles and zeros slightly away from the imaginary axis: this means that infinite peaks and nulls such as those shown in the response magnitude characteristic of figure 9.5 will not be observed in practice; and the sudden jumps shown in the phase characteristic will be somewhat smoothed out. Apart from the difficulty of designing reactive filters to account for nonideal electrical components and terminations, their main disadvantage is that their performance generally departs considerably from the ideal characteristics of figure 9.1.

Filters composed of resistors and capacitors only (or resistors and inductors only) have their poles confined to the negative real axis in the s-plane. This restriction means that it is difficult to achieve highly selective bandpass or band-stop characteristics, although resistor-capacitor filters are widely used for relatively simple filtering tasks. Although zeros may be placed anywhere in the s-plane, a

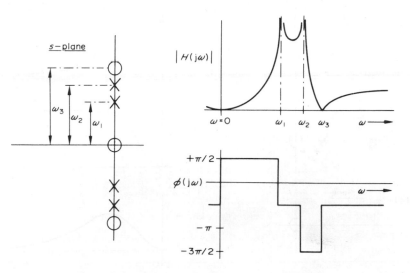

Figure 9.5   *Characteristics of a reactive bandpass filter*

common subclass of such filters (known as resistor-capacitor 'ladder networks'[36]) has zeros, like poles, confined to the real axis. Pole-zero configurations and frequency response (magnitude) characteristics for elementary low-pass, high-pass, and bandpass filters of this type are shown in figure 9.6. It is worth investigating the low-pass and high-pass characteristics a little more carefully, in order to explain the common use of the terms 'integrator' and 'differentiator' to describe them. An ideal 'integrator' would produce an output signal proportional to the time-integral of the input, and continually update it as time proceeded: the output of an ideal differentiator would be proportional at every instant to the rate of change of the input signal. We have already shown in section 3.4.4 that differentiation of a time function is equivalent to multiplying its Laplace transform by s. Similarly, integration of a time function is equivalent to dividing the Laplace transform by s. Hence we may write the transfer function of an ideal integrator and differentiator as

$$H_1(s) = 1/s, \text{ for an integrator}$$

and

$$H_2(s) = s, \text{ for a differentiator}$$

$H_1(s)$ has a single pole at the origin of the s-plane, and $H_2(s)$ a single zero at the origin. Although the pole–zero locations of figure 9.6(a) and (b) are different from these, they produce similar effects over certain frequency ranges. For example, if $\omega \gg \alpha$ the pole at $s = -\alpha$ in figure 9.6(a) has an effect very similar to one at $s = 0$: and if $\omega \ll \beta$, variations in the frequency response caused by the zero at $s = 0$ in Figure 9.6(b) far outweigh those due to the pole at $s = -\beta$. Hence if an input signal is restricted to these frequency ranges ($\omega \gg \alpha$ for the low-pass filter, $\omega \ll \beta$ for the high-pass one), the performances of these simple filters

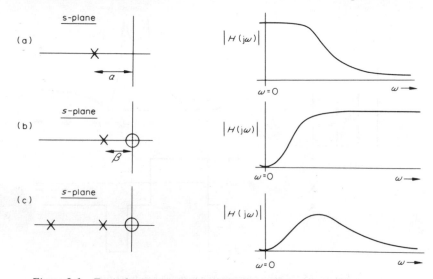

Figure 9.6   *Typical resistor-capacitor (RC) filter characteristics; (a) low-pass, (b) high-pass, and (c) bandpass*

closely match those of an ideal integrator and differentiator. It is interesting to note that in section 7.2.2 we examined the effect of a simple low-pass filter of this type on a square wave input signal (figures 7.2, 7.3, and 7.4). Indeed, figure 7.4 illustrated its effect on a square wave containing only relatively high-frequency components, and showed that the output waveform was triangular in shape. In view of the present discussion this result is unsurprising, since the time-integral of a square wave is a triangular wave having the same period. Better approximations to the ideal differentiator and integrator which operate over a larger range of input frequencies may be realised by using active filters and are widely employed in analogue computers[37].

The use of all three types of linear electric circuit element—resistors, capacitors, and inductors—enables zeros to be placed anywhere in the s-plane, and poles to be placed anywhere to the left of the imaginary axis: this allows a much more flexible approach towards filter design. As already noted, the use of active devices in a filter obviates the need for inductors, so that filters of this more general type are increasingly realised using resistors, capacitors and active elements. The modern approach towards analogue filter design involves specifying the s-plane locations of a given numbers of poles and zeros so as to approximate a desired frequency response magnitude (or phase) characteristic as closely as possible. In such design it is normal to take account of the actual electrical properties of the devices to which the filter is connected[35, 38].

This rather general approach towards approximating ideal filter characteristics may be conveniently illustrated by considering an important group of filters which are specified in terms of s-plane poles only. Suppose that the transfer function of such a filter is given by

$$H(s) = \frac{1}{(a_n s^n + a_{n-1} s^{n-1} + \ldots a_1 s + a_0)},$$

where the factors of the denominator polynomial represent the poles of $H(s)$. Its frequency response is

$$H(s)|_{s=j\omega} = H(j\omega) = \frac{1}{(a_n(j\omega)^n + a_{n-1}(j\omega)^{n-1} + \ldots a_1(j\omega) + a_0)}$$

and the square of its response magnitude is given by

$$|H(j\omega)|^2 = H(j\omega) . H^*(j\omega) = H(j\omega) . H(-j\omega)$$

$$= \left\{ \frac{1}{a_n(j\omega)^n + a_{n-1}(j\omega)^{n-1} + \ldots a_1(j\omega) + a_0} \right\}$$

$$\times \left\{ \frac{1}{a_n(-j\omega)^n + a_{n-1}(-j\omega)^{n-1} + \ldots a_1(-j\omega) + a_0} \right\}$$

The result of multiplying two denominator polynomials of this general form together is to produce a further polynomial in even powers of ω. For example, if one of the denominator polynomials is equal to $[a(j\omega)^3 + b(j\omega)^2 + c(j\omega) + d]$, the other is $\{a(-j\omega)^3 + b(-j\omega)^2 + c(-j\omega) + d\}$, and their product is $\{a^2 \omega^6 + (b^2 - 2ac)\omega^4 + (c^2 - 2bd)\omega^2 + d^2\}$. It is interesting to note that if the set of poles which gives rise to the frequency response $H(j\omega)$ are all in the left-half s-plane (as they must be if the filter is to be stable), then the poles which correspond to the function $H(-j\omega)$ are a mirror-image set lying in the right-half s-plane. Therefore if we approximate some desired squared magnitude function by choosing a suitable denominator polynomial in even powers of ω, we effectively define a pole configuration which is symmetrically disposed about the imaginary axis in the s-plane: a filter with the frequency response $H(j\omega)$ is then realised by choosing just those poles lying to the left of the axis.

Viewed in this way, the task of approximating a desired squared magnitude characteristic $|H(j\omega)|^2$ reduces to that of choosing a suitable denominator polynomial in ω^2. Suppose, for example, we wish to approximate the ideal low-pass characteristic of figure 9.7. To be effective, the denominator polynomial should have a magnitude close to $1 \cdot 0$ over the passband, and as large as possible

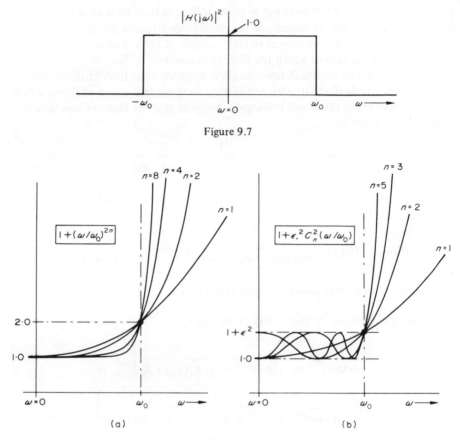

Figure 9.7

Figure 9.8

over the stopband. Figure 9.8 illustrates two types of polynomial having broadly these characteristics: part (a) shows several functions of the form

$$\left[1 + \left(\frac{\omega}{\omega_0}\right)^{2n}\right]$$

where n is a positive integer. Part (b) shows functions of the form

$$\left[1 + \epsilon^2 \cdot C_n^2\left(\frac{\omega}{\omega_0}\right)\right]$$

where C_n denotes the so-called Chebychev polynomial[38] of order n; in the range $-\omega_0 < \omega < \omega_0$, $\epsilon^2 \cdot C_n^2 (\omega/\omega_0)$ oscillates between values of 0 and ϵ^2 (for any value of n) and rises to large values outside this range. In both types of polynomial the transition from relatively small to large values in the region of $\omega = \omega_0$ is more sudden as the parameter n is increased. The use of these polynomial sets gives rise to the well-known Butterworth and Chebychev filters, which are defined by the following squared-magnitude functions

$$|H_1(j\omega)|^2 = \frac{1}{\left[1 + \left(\dfrac{\omega}{\omega_0}\right)^{2n}\right]} \qquad \text{(Butterworth)}$$

and

$$|H_2(j\omega)|^2 = \frac{1}{\left[1 + \epsilon^2 C_n^2\left(\dfrac{\omega}{\omega_0}\right)\right]} \qquad \text{(Chebychev)}.$$

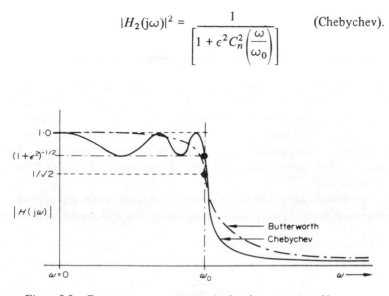

Figure 9.9   *Frequency response (magnitude) characteristics of low-pass Butterworth and Chebychev filters of fifth order*

Typical filter magnitude characteristics (for $n = 5$) are shown in figure 9.9. The magnitude of the passband 'ripple' in the Chebychev filter is controlled by the constant ϵ: if, in a particular application, more ripple is tolerated, then the transition from passband to stopband is more abrupt. n is called the order of the filter, and is equal to the number of poles in the left-hand half of the s-plane: for a given value of n, the stopband performance becomes identical for the Butterworth and Chebychev filters as $\omega \to \infty$.

The s-plane pole locations which correspond to a particular denominator polynomial may be found by making the substitution $s = j\omega$, or $\omega = s/j$. Thus in the case of the Butterworth filter, we have

$$|H(j\omega)|^2 = H(j\omega) . H(-j\omega) = \frac{1}{1 + \left(\dfrac{\omega}{\omega_0}\right)^{2n}}$$

and therefore

$$\{H(s).H(-s)\} = \cfrac{1}{1 + \left(\cfrac{s}{j\omega_0}\right)^{2n}}$$

$$= \cfrac{1}{1 + \left(\cfrac{s}{\omega_0}\right)^{2n}}, \qquad n \text{ even}$$

and

$$\cfrac{1}{1 - \left(\cfrac{s}{\omega_0}\right)^{2n}}, \qquad n \text{ odd}$$

This function has $2n$ poles equally spaced around a circle of radius ω_0 in the s-plane; as explained above, the n poles to the left of the imaginary axis define the filter. It may also be shown[38] that the poles of a Chebychev filter are arranged on an ellipse whose major axis lies along the imaginary axis in the s-plane. The greater the passband ripple, the more elongated is the ellipse. The pole locations which correspond to the filter characteristics of figure 9.9 are shown in figure 9.10.

Butterworth and Chebychev filters therefore represent particular approximations to ideal filter characteristics; other approximations, based on differen

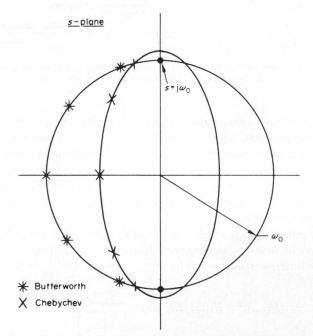

Figure 9.10   *Pole locations of the low-pass filters of figure 9.9*

polynomials, are of course possible. Another type of filter of considerable practical interest is the Thomson or Bessel filter[39], which approximates an ideal phase characteristic (phase lag proportional to frequency—that is, constant transmission delay). However the magnitude characteristics of Bessel filters are inferior to those of the types discussed above.

9.4 Digital filters

9.4.1 General

A digital filter may be realised in either 'hardware' or 'software' form. In the first, a suitable set of digital (logic) electronic circuits is interconnected to provide the essential building blocks of a digital filtering operation — storage, delay, addition/ subtraction, and multiplication by constants. Recent developments in electronics allow a complete digital filter to be constructed in integrated circuit form. A major advantage of using such special-purpose hardware, dedicated to a particular signal processing task, is speed — particularly if some of the necessary operations are performed in parallel. On the other hand, large-scale integrated circuits are very costly unless produced in high volume. The alternative, 'software', approach is to program a general-purpose digital computer as a digital filter. Since such computers are generally serial machines, which can only carry out a set of programmed instructions one at a time, operating speeds are much slower. However, powerful microcomputers are now so cheap and widely available that we must expect 'software' digital filters to become increasingly common for real-time applications requiring sampling rates up to (say) 10 kHz; and this approximate upper limit may be expected to increase as computer hardware is further developed. It must also be remembered that signals, records and data may be stored and subsequently processed 'off-line'. In such cases, a digital filter's operating speed is often not a major consideration.

Before discussing digital filters in any detail, it should be emphasised that there are two rather distinct ways in which a sampled-data signal may be filtered. The first involves taking the Discrete Fourier Transform (DFT) of the signal, probably by use of the Fast Fourier Transform (FFT) algorithm. Both the DFT and the FFT have been covered in our discussion of sampled-data signals in chapter 4. Having found the signal spectrum, the magnitudes and phases of its various frequency components may then be adjusted in accordance with the desired filter characteristics, and the filtered time-domain signal evaluated by inverse transformation — again using the FFT. In this first method, the filtering may be considered to take place in the frequency domain. It is an important and widely used approach, which allows great flexibility in the choice of filter characteristics. Since the signal spectrum is simply *multiplied* by the desired filter characteristic, it often proves faster than the equivalent time-domain *convolution*. On the other hand, its flexibility is not needed for many practical filtering tasks, and time-domain filtering, as discussed below, is often preferable. Although the principle of the FFT has been covered in section 4.3.2., its detailed implementation is a rather specialised matter, included in various other texts.[10, 20] We shall therefore not discuss it further here. However, it is worth noting that many general purpose computers include FFT programmes as part of their standard software, and that special-purpose FFT hardware is now commercially available.

The second method of implementing a digital filter, which we shall be considering in some detail in this section, is to work entirely in the time-domain. In effect, this is done by convolution of the input signal with the impulse response of the appropriate filter. Whether a time-domain or a frequency-domain filter is more appropriate depends upon a number of factors such as the storage available in the computer, the duration of the signal, the operating speed, and whether or not a 'real-time' operation is required. In such an operation, a new output sample value is calculated every time an input sample is generated. This implies digital filtering in the time-domain, because efficient use of the FFT requires the storage and processing of data in substantial blocks.

Digital filters are commonly designed by first choosing a transfer function $H(z)$ which represents a useful frequency-response characteristic. Knowing $H(z)$, it is a simple matter to derive the time-domain formula which describes the operation of the filter. This is by no means the only way of proceeding, but we start with it here because it neatly illustrates many important aspects of filter design and implementation.

A powerful way of choosing a suitable transfer function $H(z)$ for a particular filtering task is to specify a set of z-plane poles and zeros. We saw in section 4.4.2 how a sampled-data signal could be represented by such poles and zeros, and how its frequency spectrum could be inferred by considering the changes in lengths and phase of vectors drawn from the various poles and zeros to points on the unit circle (representing a series of sinusoidal frequencies). When we turn our attention to digital filters the same general concepts apply, except that we are now concerned with frequency responses rather than frequency spectra. Suppose, for example, we have a digital filter whose transfer function has the poles and zeros shown in figure 9.11. Its response at some frequency ω_1 may be found by drawing vectors from the poles and zeros to the point $z = \exp(sT) = \exp(j\omega_1 T)$, as indicated in the figure. Denoting the lengths of these vectors by a_1, a_2 and b_1, the response magnitude at frequency ω_1 is simply equal to

$$\frac{a_1 a_2}{b_1}$$

By similar arguments we may find the filter's response to a whole range of input frequencies. It is worth noting that point A in figure 9.11 represents sinusoidal frequencies $\omega = 0, 2\pi/T, 4\pi/T, \ldots$, and point B represents frequencies $\omega = \pi/T, 3\pi/T, 5\pi/T \ldots$. This means, of course, that the frequency-response characteristic of any digital filter repeats indefinitely at intervals in ω of $2\pi/T$ radians/second — as already illustrated in figure 9.2. It also means that the repetitive nature of a digital filter's frequency response exactly mirrors that of the spectrum of any sampled-data signal which it processes.

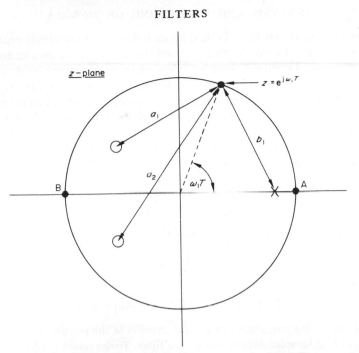

Figure 9.11   *The z-plane poles and zeros of a digital filter*

9.4.2 *Some elementary digital filters*

In order to see just how a digital filter operates, and how it may be implemented, it is helpful to start with a few simple examples and to explore the relationships between their time-domain and frequency-domain properties. In subsequent sections we will use the same general approach to examine some rather more sophisticated designs.

Figure 9.12(a) shows a very simple low-pass filter specified by a single zero at $z = -1$. At frequency $\omega = 0$, the zero vector (drawn to the point $z = 1$ on the unit

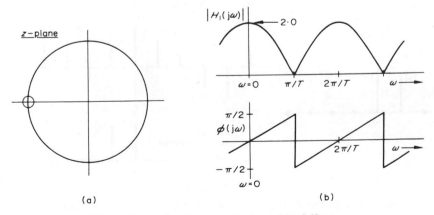

Figure 9.12   *An elementary low-pass digital filter*

circle) has a maximum length of 2·0 and hence the response magnitude must also be a maximum; as we move anticlockwise around the unit circle the zero vector becomes shorter and eventually vanishes when we reach the point $z = -1$, which corresponds to the sinusoidal frequency $\omega = \pi/T$. As the frequency further increases, the zero vector becomes longer again, reaching its maximum length of 2·0 when $\omega = 2\pi/T$; the cycle then repeats. By considering the phase angle which the zero vector makes with the positive real axis, we may also infer the filter's phase response. Magnitude and phase response are separately illustrated in figure 9.12(b). The transfer function of this elementary digital filter is

$$H_1(z) = (z + 1)$$

and equals the transform of the output signal divided by the transform of the input: if we write these as $Y(z)$ and $X(z)$ respectively, then

$$H(z) = \frac{Y(z)}{X(z)} = (z + 1)$$

or

$$Y(z) = (z + 1) \cdot X(z) = z \cdot X(z) + X(z)$$

This is, of course, the frequency-domain description of the relationship between the input and output sampled-data signals of the filter. To see how the filter may be implemented in practice, it is helpful to find its impulse response. Just as a term z^{-1} in a z-transform expression represents a unit sample at $t = T$, so a term z^1 represents a unit sample at $t = -T$. The impulse response of this particular filter is therefore as shown in figure 9.13(a). Suppose we now convolute this impulse response with a typical input signal having successive sample values . . . $x(n - 1)$, $x(n)$, $x(n + 1)$, $x(n + 2)$ . . . , as illustrated in part (b) of the figure. To calculate the output sample $y(n)$, which coincides with the input sample $x(n)$, we lay a reversed version of the impulse response below the input signal, crossmultiply, point by point, and the sum terms, giving

$$y(n) = 1 \cdot 0 \, x(n) + 1 \cdot 0 \, x(n + 1)$$

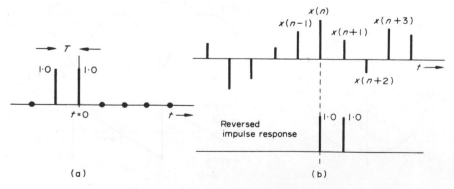

(a)   (b)

Figure 9.13   *(a) Impulse response of the low-pass filter of figure 9.12 and (b) derivation of the filter's output by convolution*

The next output value is found by moving forward one sampling period and repeating the process: each output is therefore equal to the sum of two consecutive inputs. Such an operation is easily performed using a digital computer.

An expression such as the one above, which relates input and output sample values of a digital filter, is called a 'recurrence formula'. Actually, the recurrence formula may be derived quite easily from the transfer function of a digital filter. In the above example, we have

$$Y(z) = z \, . \, X(z) + X(z)$$

from which we may write the recurrence formula by inspection as

$$y(n) = x(n + 1) + x(n)$$

In general, we may transform a term such $a_1 z^m \, . \, X(z)$ into $a_1 \, . \, x(n + m)$, or a term $a_2 z^k \, . \, Y(z)$ into a term $a_2 \, . \, y(n + k)$, where m and k are integers. The reason for this is that z may be thought of as a shift operator: multiplication by z is equivalent to a time-shift of T seconds; multiplication by z^2 is equivalent to a time-shift of $2T$ seconds, and so on. It is one of the principal attractions of the z-transform that a filter transfer function may be converted into an equivalent time-domain recurrence formula with such ease. One further word of explanation is perhaps appropriate. If, as above, we have a transfer function involving positive powers of z and hence an impulse response which begins *before* $t = 0$, any one filter output value $y(n)$ depends on one or more inputs occurring *after* the input $x(n)$. Hence the recurrence formula for $y(n)$ involves terms such as $x(n + 1)$ and $x(n + 2)$ as well as $x(n)$. Conversely, if the transfer function involves only negative powers of z, the impulse response begins after $t = 0$, and the recurrence formula involves previous inputs such as $x(n - 1), x(n - 2)$, etc. The reason for this is, of course, that convolution involves reversal of the filter's impulse response: unless this point is borne in mind, some confusion can arise over the fact that a filter transfer function involving a term or terms in positive powers of z (which represent *time-advance*) has a recurrence formula which evaluates $y(n)$ in terms of $x(n)$ and *subsequent* input values.

This very elementary example of a digital filter illustrates one or two further matters of considerable interest. Firstly, the reader may have noticed that the filter is 'unrealisable', in the sense that a given output value $y(n)$ depends upon the coincident input $x(n)$ and the next input $x(n + 1)$: it is clearly not possible for such a filter to perform a 'real-time' operation, because it can hardly anticipate the next input value (unless, of course, input samples are stored or recorded prior to processing). This difficulty always arises when a transfer function has terms in positive powers of z, caused by the number of z-plane zeros exceeding the number of poles. It may be solved by dividing $H(z)$ by a term z^m, where m equals the excess number of zeros. This is equivalent to adding m poles at the origin of the z-plane: such poles do not affect the frequency-response magnitude characteristic, but merely have the effect of delaying the output by m sampling intervals. Thus in the above example of a low-pass filter with a single z-plane zero, the unrealisable transfer function

$$H_1(z) = (z + 1)$$

may be converted into the realisable one

$$H_1'(z) = \frac{(z+1)}{z} = 1 + z^{-1}$$

by the simple expedient of including a single pole at the origin of the z-plane. The recurrence formula now becomes

$$y(n) = x(n) + x(n-1)$$

which means that any output sample is found by summing the present and previous inputs.

It is perhaps worth summarising the foregoing arguments, which apply in principle to the design and implementation of considerably more sophisticated digital filters than the low-pass type described above. Firstly, we choose a pole–zero configuration which is expected to yield the required frequency-response character-istics. This allows us to write down the filter's transfer function $H(z)$ – and, incidentally, to evaluate the frequency response exactly by substituting $e^{j\omega T}$ for z in $H(z)$. If we have specified a filter with more zeros than poles, it will be found that the filter is unrealisable, in the sense that its impulse response begins before $t = 0$: this may always be corrected by adding poles at the origin of the z-plane. Having decided upon a transfer function $H(z)$, we may with great ease convert it into a recurrence formula, which indicates how the filter may be implemented by a time-domain operation. As we have seen, the recurrence formula is essentially a description of the convolution procedure.

Before considering other simple types of filter, it is interesting to check that the recurrence formula we have derived does indeed represent a low-pass filter. Suppose our input signal is a sampled version of a steady (that is, zero-frequency) waveform. Since all input samples are equal in this case, each output (which is the sum of two inputs) will have a value twice as great as any input: thus the filter's gain to a zero-frequency input must be 2·0, as predicted by figure 9.12. Conversely, if the filter input is a sampled version of a sinusoid of frequency $\omega = \pi/T$ radians/second, successive input samples will be equal in magnitude but opposite in sign. Our recurrence formula will now give us a zero output at every sampling instant; in other words the frequency $\omega = \pi/T$ is completely rejected by the filter, as we would expect from the presence of a z-plane zero at $z = -1$. Frequencies between $\omega = 0$ and $\omega = \pi/T$ will experience a transfer magnitude somewhere between 0·0 and 2·0.

For our next example of a digital filter we investigate another low-pass characteristic, this time achieved by placing a single pole at $z = \alpha$ as in figure 9.14(a). If we make α just less than unity, the pole vector will be very small when $\omega = 0$, giving a large response; and relatively large when $\omega = \pi/T$, giving a small response. The frequency-response magnitude characteristic is shown in figure 9.14(b). The transfer function in this case is

$$H_2(z) = \frac{1}{(z-\alpha)} = \frac{Y(z)}{X(z)}$$

Hence

$$(z-\alpha) \cdot Y(z) = X(z)$$

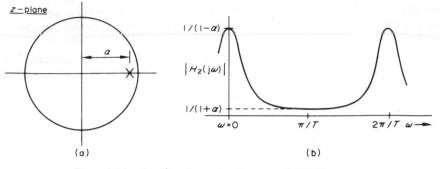

Figure 9.14   *Another elementary low-pass digital filter*

or

$$z \,.\, Y(z) - \alpha \,.\, Y(z) = X(z)$$

The corresponding recurrence formula is therefore

$$y(n + 1) - \alpha \,.\, y(n) = x(n)$$

which is equivalent to

$$y(n) - \alpha \,.\, y(n - 1) = x(n - 1)$$

or

$$y(n) = \alpha \,.\, y(n - 1) + x(n - 1)$$

This interesting result shows that in this case an output sample $y(n)$ is found from the previous output and the previous input, rather than from input samples only. A filter which calculates a new output value using one or more previous outputs is called 'recursive', and arises whenever a transfer function has poles placed other than at the origin of the z-plane. Actually, we could also realise this filter non-recursively, by writing its transfer function as

$$H_2(z) = \frac{1}{(z - \alpha)} = (z^{-1} + \alpha z^{-2} + \alpha^2 z^{-3} + \alpha^3 z^{-4} + \dots)$$

giving the recurrence formula

$$y(n) = x(n - 1) + \alpha \,.\, x(n - 2) + \alpha^2 \,.\, x(n - 3) + \alpha^3 \,.\, x(n - 4) + \dots$$

However this would hardly be sensible, since the calculation of each output now requires operations on a very large (and theoretically infinite) number of previous inputs. It is interesting to note that the use of a non-recursive recurrence formula is equivalent to a normal type of convolution procedure; indeed the various coefficients by which input samples are multiplied ($1, \alpha, \alpha^2, \alpha^3, \dots$ in the above example) are equal to the values of successive terms in the filter's impulse response. On the other hand, the multiplier coefficients of a recursive recurrence formula bear no simple or obvious relation to the impulse response. As we have hinted above, a recursive filter is normally far preferable to a non-recursive one with similar frequency-response characteristics, because it is so much more economical in terms of computational effort.

Bearing in mind that when applied to a digital filter the term 'high-pass' implies strong transmission of frequencies in the neighbourhood of $\omega = \pi/T$, we may modify the above pole-zero locations to produce a simple form of high-pass filter. First, let us place a single pole at $z = -\alpha$, where α is just less than unity: if we also require complete rejection of zero-frequency components, we place a zero at $z = 1$, as shown in figure 9.15(a). Part (b) of the figure shows the magnitude response characteristic of the filter. The high-pass transfer function is

$$H_3(z) = \frac{(z-1)}{(z+\alpha)} = \frac{Y(z)}{X(z)}$$

hence

$$(z + \alpha) \cdot Y(z) = (z - 1) \cdot X(z)$$

and the recurrence formula is therefore

$$y(n + 1) + \alpha \cdot y(n) = x(n + 1) - x(n)$$

or

$$y(n) + \alpha \cdot y(n - 1) = x(n) - x(n - 1)$$

giving

$$y(n) = -\alpha \cdot y(n - 1) + x(n) - x(n - 1)$$

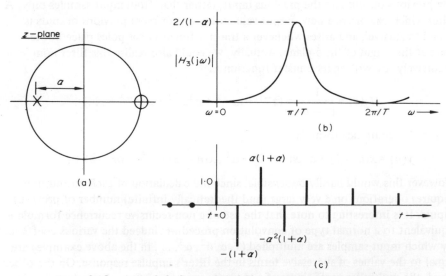

Figure 9.15   *A simple high-pass digital filter. (a) pole-zero configuration,*
(b) magnitude response characteristic, and (c) impulse response

$H_3(z)$ could also be expressed in the form of a numerator polynomial in z, from which the filter's impulse response could be obtained by inspection. Another, and perhaps simpler, way to find the impulse response is to assume the filter input to consist of a single unit-height sample and to evaluate the output term by term, using the recurrence formula. For example, if we assume that $x(n) = 1$, that all other input samples are zero, and that the output is also zero prior to the arrival of $x(n)$ at the input, successive outputs are given by

$$y(n) = -\alpha(0) + 1 - 0 = 1$$

$$y(n + 1) = -\alpha \cdot y(n) + x(n + 1) - x(n) = -\alpha + 0 - 1$$

$$= -(1 + \alpha)$$

further outputs are unaffected by the term $x(n)$ so that

$$y(n + 2) = -\alpha \cdot y(n + 1) = \alpha(1 + \alpha)$$

$$y(n + 3) = -\alpha \cdot y(n + 2) = -\alpha^2(1 + \alpha)$$

$$y(n + 4) = -\alpha \cdot y(n + 3) = \alpha^3(1 + \alpha), \text{ etc.}$$

The first nine terms of the impulse response are shown in figure 9.15(c). Once again, this response defines the coefficients by which a series of input samples would have to be multiplied in a non-recursive version of the filter. Even if we were to ignore very small impulse response terms, it is clear that the non-recursive version would require a great deal more computational effort than that specified by the recursive recurrence formula.

Our final example of a simple digital filter has the pole—zero configuration of figure 9.16(a). The complex conjugate pole-pair placed on a circle of radius r (where r is close to unity) gives rise to a bandpass characteristic centred on ω_0 radians/second: zero-frequency and high-frequency rejection are provided by the zeros at $z = 1$ and $z = -1$. The corresponding response magnitude characteristic is also shown in the figure. If, as is quite often convenient, we express the pole positions in polar

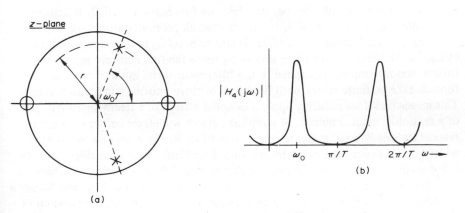

Figure 9.16   *A bandpass digital filter*

coordinates, the transfer function is

$$H_4(z) = \frac{(z+1)(z-1)}{[z - r.\exp(j\omega_0 T)]\,[z - r.\exp(-j\omega_0 T)]}$$

$$= \frac{z^2 - 1}{z^2 - rz\,[\exp(j\omega_0 T) + \exp(-j\omega_0 T)] + r^2}$$

$$= \frac{z^2 - 1}{z^2 - 2rz\,\cos\,\omega_0 T + r^2}$$

From this expression we may derive the recurrence formula

$$y(n) = 2r\cos\,\omega_0 T.\,y(n-1) - r^2.\,y(n-2) + x(n) - x(n-2)$$

Two general points deserve mention. Firstly, transfer function poles must always be placed inside the unit circle if instability is to be avoided. Instability may be simply demonstrated by considering once again the function

$$H(z) = \frac{1}{z - \alpha}$$

except that this time we make $\alpha > 1$. As we have already shown, such a transfer function may be written as

$$H(z) = z^{-1} + \alpha z^{-2} + \alpha^2 z^{-3} + \alpha^3 z^{-4} + \dots$$

The filter's impulse response therefore has successive terms $1, \alpha, \alpha^2, \alpha^3, \alpha^4 \dots$ etc., and if $\alpha > 1$ it is clear that this response will continue to grow without limit. When the response to a finite, time-limited input grows indefinitely, the filter is clearly unstable — and unusable. The second point concerns the start-up of a digital filter. Suppose we have a recurrence formula of the general form

$$y(n) = a_1 y(n-1) + a_2.\,y(n-2) + \dots + b_0 x(n) + b_1.\,x(n-1)$$

$$+ b_2.\,x(n-2) + \dots$$

The first non-zero output value occurs when the first non-zero sample is delivered to the filter's input. (It is sensible to assume that all previous outputs are zero: any other assumption implies that the filter is still responding to some previous input.) Hence the first non-zero output is simply b_0 times the first non-zero input. As further non-zero inputs are applied to the filter more terms in the recurrence formula take on finite values, until finally all of them must be taken into account. This means that if we suddenly apply a sampled version of a steady input level, or of a sinusoidal input, a number of sampling periods will elapse before the output reaches its 'steady-state' response: there is, in other words, a start-up transient, which is just as essential a characteristic of a digital filter as of an analogue one. It is important to realise that when we speak of a digital filter 'rejecting' a certain frequency we mean that its steady-state response is zero. But it will always display a transient response when an input at this frequency is first applied; the duration of the transient will match that of the filter's impulse response.

9.4.3 Hardware and software implementation

We should be clear about the method of implementing a digital filter in either hardware or software form. Consider as an example the bandpass design already illustrated in figure 9.16 and suppose that, in a particular application, we choose the following parameter values:

sampling frequency = 1 kHz, that is, $T = 0.001$ s

centre frequency = 80 Hz, giving $\omega_0 T = 0.16$ radian or $28.8°$.

The recurrence formula becomes

$$y(n) = 1.6650\, y(n-1) - 0.9025\, y(n-2) + x(n) - x(n-2)$$

The block diagram of figure 9.17 illustrates a hardware implementation of this filter. Let us assume that the input signal starts off in analogue form, as on the left of the diagram. It is first sampled at 1 kHz using an analogue-to-digital converter (ADC). This produces successive input sample values $x(n)$, normally in a coded binary format. Two delay units are needed to provide the term $x(n-2)$. The output from the adder is $y(n)$, which is fed back via further delay and multiplier elements to give the recursive terms in the recurrence formula. $y(n)$ may be converted back from a binary code to the equivalent analogue voltage using a digital-to-analogue converter (DAC). In terms of digital hardware, the multiplication operations are the most complicated, and normally limit the speed of operation.

Software implementation is quite straightforward, and is illustrated below using a BASIC program. Let us assume off-line operation, and that 1000 input signal values have been loaded into a storage array X. The filtered output values are to be stored in array Y. A suitable series of instructions would be:

```
DIM X(1000), Y(1000)

Y(1) = 0

Y(2) = 0

FOR N = 3 to 1000

Y(N) = 1.6650 * Y(N - 1) - 0.9025 * Y(N - 2) + X(N) - X(N - 2)

NEXT N

STOP
```

The essential feature is an iterative loop, in which the recurrence formula is used over and over again to calculate successive filter output values. As previously noted, there will be a start-up transient in the response, and care must be taken to ensure that the filter is suitably initialised and that storage arrays are not overrun. If a software filter is to run in real-time, its maximum operating speed can normally be substantially increased by programming in a language which is closer to the digital 'language' of the computer itself, rather than in a high-level language such as BASIC or FORTRAN. Instructions will also have to be included in the program loop to operate any ADCs and DACs.

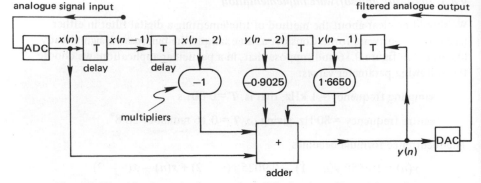

Figure 9.17   *A hardware implementation of the bandpass filter of figure 9.16*

Many of the basic concepts, and much of the terminology, of digital filters have now been covered. Although simple designs such as those already described can be valuable in a variety of relatively undemanding applications, a wide range of more advanced design methods is available. In order to consider a few of these in more detail, it is helpful to divide digital filters into two major classes, depending upon whether they have finite impulse responses (FIRs) or infinite impulse responses (IIRs).

9.4.4 Filters with finite impulse responses (FIRs)

9.4.4.1 Introduction

The form of filter shown in figure 9.18, which has m delay stages, $(m + 1)$ positive or negative multipliers, and an adder or summing junction, is often referred to as a digital transversal filter. It is non-recursive, with a time-domain recurrence formula given by:

$$y(n) = a_0x(n) + a_1x(n - 1) + a_2x(n - 2) + \ldots + a_mx(n - m)$$

$$= \sum_{i=0}^{m} a_ix(n - i)$$

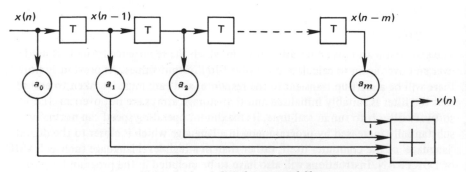

Figure 9.18   *A digital transversal filter*

If we consider a single, unit-valued, input sample to this filter, it will clearly generate a sequence of sample values $a_0, a_1, a_2 \ldots a_m$ at the output. Therefore the filter's impulse response is just made up of the multiplier coefficient sequence a_0 to a_m, and is finite in duration. The art of designing such a filter is to specify the minimum number of delay and multiplier elements to achieve an acceptable performance. The first of the low-pass filters discussed in section 9.4.2 is of this type, but with only two unit multipliers; these provide a very elementary low-pass function. Most useful FIR filters require between (say) 15 and 150 multipliers, and have an equivalent number of zeros (but no poles) in their transfer functions. Their design can hardly therefore be based upon a suitable choice of z-plane zero configuration, and other methods must be used. Although non-recursive FIR filters usually require many more delay and multiplier elements than recursive filters of comparable performance, they have two major compensating advantages:

(a) An impulse response which is finite in duration can also be symmetrical in form. This produces a pure linear-phase characteristic[28] (equivalent to imposing the same pure time-delay on all frequency components in the input signal). There is, in other words, no phase distortion. FIR filters do not *have* to be linear-phase, but most practical designs take advantage of this possibility, which is not available in analogue filters based upon lumped circuit elements such as resistors, capacitors, and inductors.
(b) Non-recursive designs are inherently stable, since they do not involve feedback from output to input. There is no risk that inaccurate specification of one or more of the multipliers might lead to instability.

9.4.4.2 *The moving-average filter*

We start by considering a simple form of digital transversal filter in which all the multipliers are equal. This is often called a moving-average filter. Suppose, for example, we use 19 delay elements and 20 multipliers all equal to 1/20 or 0·05. This gives the impulse response shown in figure 9.19(a). Convolution of this response with an input signal yield output sample values, each of which is the *average* of 20 consecutive inputs. Part (b) of the figure shows a typical case: the input sequence (shown joined together by a solid line for clarity) is assumed to have its first non-zero value at $t = 0$, and displays a steady downward trend together with superimposed random high-frequency fluctuations. These fluctuations might represent observation or measurement errors. The filter's output (shown as a series of dots) has a start-up transient between $t = 0$ and $t = 19T$, which corresponds to the duration of its impulse response; thereafter it transmits the slow trend of the input but greatly reduces the rapid fluctuations. This smoothing, or averaging, action is equivalent to low-pass filtering.

The pure delay imposed by a linear-phase filter may be shown[28] to equal half the duration of its symmetrical impulse response; in this case, the delay is $9\frac{1}{2}T$. If the output waveform in figure 9.19(b) is advanced in time by this amount, it is seen to follow the slow trend of the input very closely, with a gain of about unity. The filter's low-frequency gain of unity is confirmed by its frequency-response

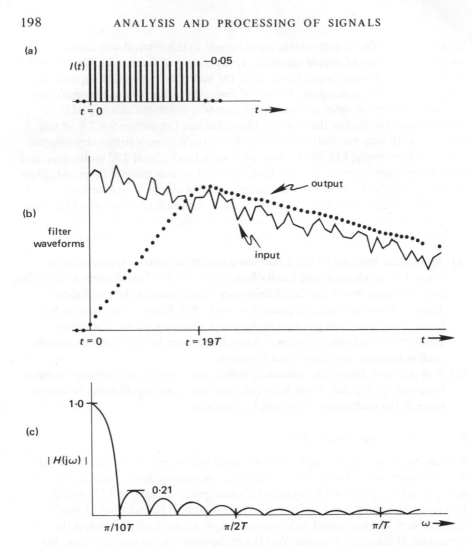

Figure 9.19   *A moving-average filter: (a) impulse response, (b) typical input and output waveforms, and (c) frequency-response magnitude characteristic*

magnitude characteristic, shown in figure 9.19(c). As expected, this is broadly of the low-pass type, although the presence of substantial sidelobes makes it rather non-ideal. Nevertheless, moving-average filters are widely used for undemanding low-pass filtering tasks. Note that, since all multiplier coefficients are equal, a *single* multiplier at the output of the adder would have the same effect as the many individual ones on the input side. Alternatively, for even greater simplicity, the multipliers may be omitted altogether. This merely alters the output by a scale factor, but does not otherwise affect filter performance. In this example, we have illustrated a moving-average design with 20 multipliers, or tapping points, giving a frequency response with its first transmission null at $1/20T$ Hz ($\omega = \pi/10T$). Any other number of multipliers may, of course, be specified: the greater the number, the more pronounced the low-pass filtering action, and the more closely the frequency-response magnitude characteristic approximates a sin x/x form,[12] in the range $-\pi/T < \omega < \pi/T$.

9.4.4.3 *FIR designs based on window functions*

Digital filter design often starts from a frequency-response specification. In principle, the equivalent impulse response may then be found by inverse discrete Fourier transformation, and used to define the multipliers of a digital transversal filter. Unfortunately, the impulse response will often contain an unacceptably large — or even infinite — number of terms. For example, if we begin with an ideal low-pass characteristic (as previously illustrated in figure 9.2), the impulse response will be of sampled sin x/x form, with infinitely long 'tails'. One possibility is simply to ignore the smaller terms in the tails. Such truncation of the impulse response is, however, equivalent to multiplying it by a rectangular pulse or 'observation window' (see section 8.3.1) and causes spreading of the frequency-domain characteristic. This manifests itself in an undesirable broadening of the transition from passband to stopband, with a tendency to produce unwanted overshoot — the so-called Gibbs' phenomenon. The net result is often an unsatisfactory approximation to the desired frequency response. An alternative solution to the truncation problem is to select a more effective window function, such as a Hamming, Hanning, or Kaiser window.

The approach is illustrated in figure 9.20, which takes as its starting point an ideal low-pass filter characteristic $G_1(j\omega)$ with cut-off frequency $\omega_0 = \pi/4T$. Its impulse response is most simply found by initially assuming $G_1(j\omega)$ to be purely real (that is, imposing no phase shift at any frequency). The computed impulse response $I_1(t)$ is then symmetrical about $t = 0$, and of sampled sin x/x form, as shown in part (b) of the figure. We next decide how many impulse response terms, and hence transversal multipliers, we can accommodate in the final design, and multiply by a window function with this number of terms. For example, we could use a Hamming window defined as

$$w_n = 0{\cdot}54 + 0{\cdot}46 \cos n\pi/N, \quad -N < n < N$$

$$= 0 \text{ elsewhere}$$

This has $(2N - 1)$ terms. Suppose we can accept 21 transversal multipliers, so that $N = 11$. The window, and its product with the original impulse response, are then as shown in parts (c) and (d) of the figure respectively. It now only remains to shift the impulse response forward so that it begins at $t = 0$, as in part (e). This makes it physically realisable and converts the original zero-phase filter into a pure linear-phase one. The resulting frequency-response magnitude characteristic is shown at the bottom of figure 9.20, together with that obtained using a 21-term rectangular window. Note that the Hamming window achieves far better stopband performance, with sidelobe levels considerably less than 1% of the main lobe (they are hardly visible on a linear plot), at the expense of a slightly worse initial cut-off slope. The

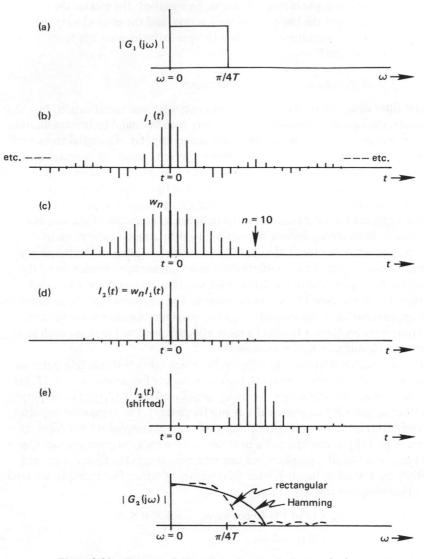

Figure 9.20   *Transversal filter design by the window method*

response would improve if a wider window was specified; but that would, of course, require more multipliers.

Other window functions may be used. The Hanning window, sometimes referred to as a 'raised cosine bell' function, is defined as

$$v_n = 0.5 + 0.5 \cos n\pi/N, \quad -N < n < N$$

and gives a slightly different compromise between passband and stopband performance. The Kaiser window, defined in terms of Bessel functions, offers excellent sidelobe suppression, at the expense of a slightly inferior initial cut-off slope. The use of these and other windows is covered in a number of texts.[10, 32, 41] Note that they are all symmetrical in form and therefore preserve the linear phase properties of a symmetrical impulse response sequence.

9.4.4.4 Frequency-sampling filters

Although it is sometimes implied that *all* FIR filters are non-recursive, this is not the case. It is true that they *may* always be realised non-recursively, but in some cases an equivalent recursive operation is not only possible, but much more economic. The best-known example of this is the so-called 'frequency-sampling' filter. To understand this important type of design, it is helpful to start by considering a digital resonator having a complex conjugate pole-pair on the unit circle in the z-plane, together with a second-order zero at the origin. Its transfer function is given by

$$H(z) = \frac{z^2}{(z - e^{j\theta})(z - e^{-j\theta})} = \frac{z^2}{z^2 - 2\cos\theta \cdot z + 1} = \frac{Y(z)}{X(z)}$$

and its time-domain recurrence formula is

$$y(n) = 2\cos\theta \cdot y(n-1) - y(n-2) + x(n)$$

As a simple example, let $\theta = 60°$, so that $\cos\theta = 0.5$. Then

$$y(n) = y(n-1) - y(n-2) + x(n)$$

Since the pole-pair is placed on the margin of stability, we expect the impulse response of this resonator to continue forever, neither increasing nor decreasing in amplitude. This is confirmed by figure 9.21(a), which shows a continuing oscillation at the (sampled) sinusoidal frequency corresponding to the pole positions.

As it stands, such a resonator is not a useful filter. However, its impulse response can be made finite by cascading it with a very simple form of digital transversal filter, known as a 'comb' filter. This is illustrated in figure 9.21(b). An impulse input to the transversal part, which has only two tapping points, or 'taps', gives rise to one positive and one negative output sample, separated by mT seconds. The first of these excites the digital resonator, which starts to generate its characteristic oscillation; the second brings it to a halt mT seconds later. This combination of transversal and resonator sections therefore produces a recursive FIR digital filter.

It is instructive to consider z-plane pole–zero locations. The transversal part has the transfer function

$$H_1(z) = 1 - z^{-m} = \frac{z^m - 1}{z^m}$$

giving m zeros uniformly spaced around the unit circle. These produce the 'comb' filter frequency characteristic shown in figure 9.21(c), for the case when $m = 24$. Since the resonator has a complex conjugate pole-pair, the overall pole–zero configuration is as shown in part (d) of the figure. We see that the poles of the

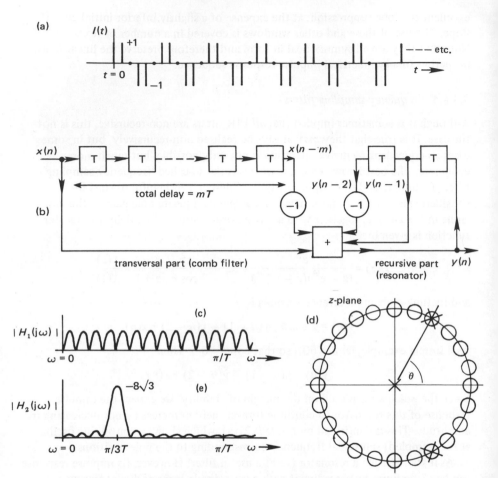

Figure 9.21 *Illustration of the basis of the frequency-sampling technique*

resonator are exactly cancelled by two of the transversal zeros, giving an overall filter which has, in effect, only z-plane zeros. This must, of course, be the case if its impulse response is finite. At first sight it might appear rather pointless to use a resonator, only to have its poles cancelled; but in fact it provides a more economic filter. The recursive recurrence formula is

$$y(n) = y(n-1) - y(n-2) + x(n) - x(n-24) \quad \text{for } m = 24$$

which requires one addition and two subtractions. But since the impulse response continues for 24 sampling periods, with eight values of +1 and eight of −1, a non-recursive realisation would need many more additions and subtractions. Clearly, the advantage of recursive operation becomes even greater if the parameter m is increased.

Figure 9.21(e) shows that the inclusion of the resonator converts the comb filter response into a bandpass characteristic. If the parameter m is increased, the width of the main passband reduces, and the characteristic becomes closer to a $\sin x/x$ function. Such a filter might be valuable in its own right for simple applications; more importantly, as described below, it forms the basic building block of a complete frequency-sampling filter.

It is interesting to note, in passing, that the moving-average filter discussed earlier (see figure 9.19) may also be implemented recursively. The appropriate 'resonator' has a single pole at $z = 1$, and an impulse response made up of a continuing series of +1 values. The recursive recurrence formula is

$$y(n) = y(n-1) + 0.05x(n) - 0.05x(n-20)$$

which is clearly much more economic than its non-recursive, transversal, equivalent.

We now return to the frequency-sampling filter. Suppose we require a digital filter with the frequency-response magnitude characteristic shown in figure 9.22(a). We first sample it, as in part (b) of the figure. The required response may now be built up by superposing a set of $\sin x/x$ functions, each weighted according to one of these

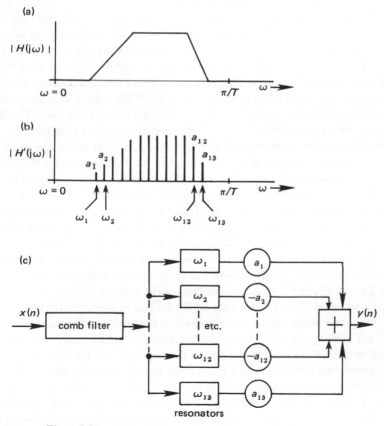

Figure 9.22 *Design of a complete frequency-sampling filter*

sample values, and arranged around it. This idea was developed in section 8.2.2, when reconstituting a continuous waveform from its sampled version — although we are now, of course, reconstituting a frequency function rather than a time function. Each of the required $\sin x/x$ responses is provided by a comb-filter-plus-resonator combination of the type shown in figure 9.21(b). Fortunately, a single comb filter can feed all the required resonators in parallel, giving the overall filter structure shown in figure 9.22(c). Note that alternate weighting coefficients must be inverted, because there is a phase reversal between the outputs of adjacent resonators. In principle, this frequency-sampling technique offers a highly flexible method of building up a desired filter characteristic using a number of subfilters with $\sin x/x$ responses. A fuller account of the method is given in most texts on digital filters.[10,11,46]

9.4.4.5 *Recursive integer-multiplier designs*

So far we have made no distinction between integer and decimal arithmetic. In fact, a digital filter in which signals and multipliers are all represented in integer form will generally be simpler to implement, and much faster in operation. Unfortunately, most digital filter design techniques yield decimal multipliers. In the case of a transversal filter designed, say, using the window function approach (section 9.4.4.3), the multipliers may be converted into integers by scaling up and rounding to the nearest whole number (assuming, of course, that the errors introduced by such 'quantisation' are acceptable). Although this cannot generally be done with the feedback terms of a recursive filter, there is a class of recursive filters which inherently possesses small integer multipliers.[42] Such filters have ideal linear-phase responses although their magnitude characteristics are often rather poor. For this reason, many texts — particularly those with a communications engineering emphasis — do not discuss them. But they are of considerable interest in other signal processing applications where operating speed is more important than a tightly specified magnitude characteristic, and may readily be programmed in machine code on a wide range of mini and microcomputers.[43]

The simple bandpass filter already illustrated in figure 9.21 is of this type. Note that the sample values in its impulse response, and the multipliers in its recursive recurrence formula, are exactly ±1. If the input signal values are expressed as integers, the complete filtering operation can be performed in integer arithmetic. However, the multiplier associated with the recurrence formula term $y(n - 1)$ is equal to $2 \cos \theta$, where θ defines the position of the resonator's complex pole pair. If $2 \cos \theta$ is to be an integer, θ can only be $0°$, $60°$ (as in figure 9.21(d)), $90°$, $120°$ or $180°$. Only five centre frequencies are therefore available in the range $0 < \omega < \pi/T$. This means that a complete frequency-sampling filter, such as that illustrated in figure 9.22, cannot be realised in integer arithmetic, since most, if not all, of its resonators will have their z-plane poles elsewhere on the unit circle.

The combination of a simple comb filter and digital resonator to produce recursive, integer-multiplier, designs therefore involves two important restrictions: limited choice of centre frequency; and rather poor amplitude characteristics, with substantial sidelobes. In spite of these drawbacks, they can be valuable for simple filtering tasks, including fast real-time operation on mini and microcomputers. One way of improving the sidelobe performance is to square (or even cube) the filter's transfer function. This is equivalent to passing the input signal through the filter twice (or three times). As an example consider the high-pass transfer function

$$H_1(z) = \frac{z^8 - 1}{z^7(z + 1)}$$

The pole at $z = -1$, corresponding to $\theta = 180°$, is provided by the resonator; the eight zeros, equally spaced around the unit circle, by the comb filter. If we square $H_1(z)$, all these poles and zeros become second-order, and the transfer function and recurrence formula are changed to

$$H_2(z) = \frac{(z^8 - 1)^2}{z^{14}(z + 1)^2} = \frac{1 - 2z^{-8} + z^{-16}}{1 + 2z^{-1} + z^{-2}}$$

and $y(n) = -2y(n - 1) - y(n - 2) + x(n) - 2x(n - 8) + x(n - 16)$. The increased number of terms in this formula is the price to be paid for a reduction in level of the first sidelobe from about 21% to about 4·5% of the main lobe. The impulse and frequency responses of this design are illustrated in figure 9.23.

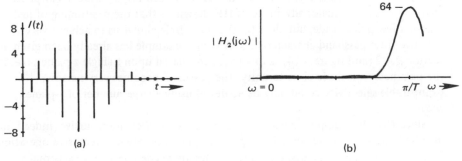

Figure 9.23   An integer-multiplier high-pass filter: (a) impulse response, and (b) frequency-response magnitude characteristic

It is interesting to note that a composite filter, built up by combining (adding) the outputs of two or more linear-phase filters with the same transmission delay, has a frequency response equal to the algebraic sum of the individual responses. The frequency-sampling technique described in section 9.4.4.4 relies upon this fact. It also allows us to design an elementary narrow bandstop ('notch') filter by subtracting the output of an integer-coefficient bandpass filter from that of an 'all-pass' network (constant gain and transmission delay). The resulting design is sufficiently fast in operation to allow, for example, the real-time rejection of mains frequency (50 Hz) interference from a biomedical signal such as the electrocardiogram, using a simple mini or microcomputer.[43]

The performance of recursive, integer-multiplier designs may be greatly improved by replacing the comb filter with a general transversal filter having a much larger number of taps. In a study based on the requirements of data transmission systems,[44] it has been shown that a transversal filter plus resonator is often much more economic than a transversal filter alone. The reason for this is that, when a resonator is used, the transversal coefficients may be reduced to small integer values (typically between, say, ±4) with little loss of overall accuracy. Although the number of transversal taps may be quite large — perhaps up to 100 to 150 if a narrowband filter characteristic is specified — the use of small integer multipliers again makes it simple to program such filters on a basic microcomputer. Software filters of this type seem likely to find increasing use in a variety of real-time applications requiring sampling rates up to a few kilohertz.[45]

9.4.5 Filters with infinite impulse responses (IIRs)

9.4.5.1 Introduction

Any digital filter specified in terms of one or more z-plane poles has an infinite impulse response (IIR). Actually, this statement should be slightly qualified by saying *uncancelled* poles, because, as we have seen in the previous section, certain types of FIR filter have z-plane poles which are cancelled by coincident zeros. The IIR of any realisable filter cannot be symmetrical in form, because it cannot start before $t = 0$. Unfortunately, therefore, no IIR filter can display pure linear-phase characteristics. The major advantage of IIR designs is that the positioning of just one or a few poles inside, but close to, the unit circle allows us to achieve very selective filter passband characteristics. A simple example has already been given in section 9.4.2 (and figure 9.14): a low-pass filter based upon a single z-plane pole. If the parameter α is made close to unity, the passband becomes very narrow. Comparable selectivity could only be achieved using a large number of z-plane zeros.

Since IIR filters employ uncancelled z-plane poles, they are recursive; indeed, no infinite impulse response could be implemented in a purely non-recursive operation. Since, in most practical cases, only a few poles are needed to achieve acceptable performance, the recurrence formulae of IIR filters normally involve relatively few terms; this contrasts with FIR transversal filters where up to 100 or 150 terms may be needed. However, the recursive multipliers of an IIR design are generally decimal numbers, needing specification with considerable accuracy (typically 3 to 6 decimal places) if the required frequency response is to be achieved, and instability avoided. There are therefore both advantages and disadvantages in IIR, as opposed to FIR, designs.

Many practical IIR filters are based upon analogue equivalents. The reason for this is largely historical: an extensive theory of analogue filters has grown up over the last half-century, and it was natural that, in the much more recent development of digital filters, equivalent designs should be sought. Since analogue filters based upon lumped circuit elements (resistors, capacitors, inductors, amplifiers) have infinite impulse responses, this approach leads naturally to digital filters of the IIR type. Essentially, the problem is to find suitable transformations for mapping the s-plane poles and zeros of an analogue filter into the z-plane. Valuable though this approach is for deriving digital filters with, for example, Butterworth or Chebychev

characteristics (see section 9.3.2 and figure 9.9), it must be remembered that some
of the constraints of analogue filter design do not apply to the digital case. For
example, the linear-phase FIR designs discussed in the previous section do not have
direct lumped-element analogue equivalents. In addition, it is quite possible to
choose a suitable z-plane pole—zero configuration for a filter, without direct
reference to analogue designs. Some examples are given in the following section.

9.4.5.2 *Designs based upon choice of z-plane poles and zeros*

In section 9.4.2, we discussed some elementary digital filters based upon z-plane
poles: a low-pass and a high-pass design, each with a single real pole; and a bandpass
filter with a complex conjugate pole pair. Simple designs of this type can be
produced without reference to equivalent analogue filters and are often useful in
undemanding applications. A little practice on the part of the designer can lead to
rather more complicated pole—zero configurations, again chosen on an essentially
empirical basis. To take a straightforward example, we might require an elementary
bandpass function centred on some frequency ω_0, which rejected any steady (d.c.)
level in the input signal, and also any fluctuations at, or close to, the frequency
$\omega = \omega_1$. The pole—zero configuration of figure 9.24(a) could be used.

Another requirement might be the design of a 'bank' of bandpass filters having
the same peak gain but different centre frequencies. This may be achieved if each
filter has a complex conjugate pole pair and a zero positioned as in figure 9.24(b).
The zero lies at the intersection of the real axis with a line joining the two poles.

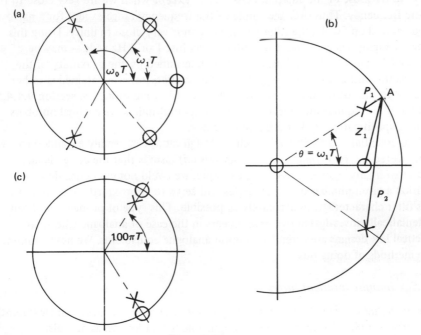

Figure 9.24

(An additional zero may be placed at the origin to minimise the filter's delay.) At the centre frequency, in this example ω_1, the peak gain G is given in terms of vectors drawn to point A on the unit circle

$$G = \frac{Z_1}{P_1 P_2}$$

Providing the required bandwidth is small, the poles will be very close to the unit circle. In this case vector Z_1 is almost exactly half vector P_2. Hence, to a very good approximation

$$G = \frac{1}{2P_1}$$

which is independent of the centre frequency chosen. Other members of the filter bank may be derived by simply changing the angle θ.

As a final example, suppose we need a narrow bandstop, or 'notch', filter for rejecting mains-frequency interference at 50 Hz ($\omega = 100\pi$) from a wideband recorded signal. The most obvious possibility would be to place a complex conjugate zero pair on the unit circle, at points corresponding to 50 Hz. Unfortunately this does not yield a sharp notch, so that there would be substantial reduction of frequency components well away from 50 Hz. A far better solution is to include a complex conjugate pole pair close to the zeros, as shown in figure 9.24(c). If we consider vectors drawn from the poles and zeros to points around the unit circle, it is clear that changes in length and direction of each pole vector are almost exactly matched by those of the adjacent zero vector, *except* when we are very close to the centre frequency. Thus over the whole of the frequency range except for a narrow band centred on 50 Hz, the filter's transmission is very close to unity. Using this design, a highly selective rejection notch — say just 1 or 2 Hz wide — may be obtained.[40] The notch width is, of course, determined by the proximity of the poles to the unit circle. The frequency-response characteristic obtained is rather superior to that of the integer-coefficient notch filter mentioned in section 9.4.4.5, and the filter requires much less storage; but its multipliers are decimal numbers needing specification to 4 or 5 figure accuracy.

This essentially empirical approach to design cannot be readily extended to more complicated filters. Perhaps the most serious criticism is that the design is not *optimised* in any rigorous sense. For example, we could not easily decide how to position a given number of z-plane poles and zeros to approximate a desired low-pass filter characteristic as accurately as possible. This type of problem has been systematically investigated over many years in the case of analogue filters, so practical IIR designs are often based upon analogue equivalents. We next examine two methods of doing this.

9.4.5.3 *Impulse invariant filters*

One of the most straightforward techniques (at least from a conceptual viewpoint) for deriving a digital filter from an analogue one is that known as 'impulse invariance'. This consists of using a sampled version of the impulse response of the

analogue filter to define the impulse response of a digital one. The resulting
relationships between the responses of analogue and digital versions of a typical
low-pass filter are illustrated in figure 9.25. Since sampling a time function has the
effect of repeating the corresponding frequency function indefinitely at intervals in
ω of $2\pi/T$, the digital filter will have a frequency response equal to a repetitive
version of that of the analogue filter. This points immediately to a potential
disadvantage of the impulse invariance approach: the digital filter's frequency
response in the range $0 < \omega < \pi/T$ may not be a good approximation to that of the
analogue filter in the same range, because of aliasing (see section 8.2.1). This may
be serious, particularly in the region of $\omega = \pi/T$, if the analogue filter transmits
significantly at this or higher frequencies. For this reason, the technique cannot be
used for high-pass filters.

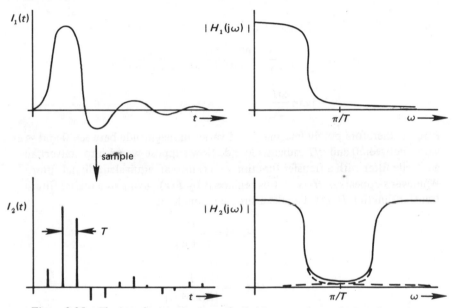

Figure 9.25   *The impulse invariance method. Above are shown the impulse and
frequency responses of an analogue filter; below, those of its digital equivalent.
$I_2(t)$ is a sampled version of $I_1(t)$*

Assuming that any aliasing is acceptable, sample values of the impulse response
of the analogue filter may be used directly to define the transversal multipliers of a
non-recursive digital filter. This is straightforward in principle, but in practice may
require a large number of terms, leading to an uneconomic filter. A better solution —
which falls beyond the scope of this chapter — is to develop a recursive design which
also displays the property of impulse invariance.[46]

9.4.5.4 *The bilinear transformation method*

The bilinear transformation method is a widely used, and very effective way of deriving a digital counterpart of an analogue filter. It avoids the problem of aliasing inherent in the impulse invariance method, at the expense of some distortion of the frequency scale; however, this latter effect is not generally a significant disadvantage. The bilinear transformation is an example of a so-called 'frequency transformation'. The mathematical background to such transformations (which should be distinguished from time-to-frequency transformations such as the Fourier, Laplace, or z-transforms) is rather complicated, although the basis of the approach may be clarified by an example. Consider first the function

$$F(z) = \frac{(z-1)}{(z+1)}, \text{ where } z = e^{sT}$$

This is 'bilinear' in the sense that both its numerator and denominator polynomials are linear in the variable z. In order to explain the value of this function in converting an analogue filter into a digital equivalent, we need to evaluate its spectrum. This is found by putting $s = j\omega$, or $z = e^{j\omega T}$, which gives

$$F(j\omega) = \frac{e^{j\omega T} - 1}{e^{j\omega T} + 1} = \frac{e^{j\omega T/2}(e^{j\omega T/2} - e^{-j\omega T/2})}{e^{j\omega T/2}(e^{j\omega T/2} + e^{-j\omega T/2})}$$

$$= j \tan \frac{\omega T}{2}$$

$F(j\omega)$ is therefore purely imaginary and varies in magnitude between 0 and ∞ as ω varies between 0 and π/T radians/second. Now suppose we wish to convert an analogue filter with a transfer function $H_1(s)$ into an equivalent digital filter. Wherever s appears in $H_1(s)$, let us replace it by $F(z)$, giving us a digital filter transfer function $H_2(z)$. To take a simple example, if

$$H_1(s) = \frac{1}{(s + \alpha)}$$

then

$$H_2(z) = \frac{1}{[F(z) + \alpha]}$$

The frequency responses of the two filters are given by

$$H_1(j\omega) = \frac{1}{(j\omega + \alpha)}$$

and

$$H_2(j\omega) = \frac{1}{[F(j\omega) + \alpha]} = \frac{1}{[j \tan \omega T/2 + \alpha]}$$

If ω varies between 0 and ∞, the term $(j\omega)$ in $H_1(j\omega)$ clearly varies between j0 and j∞; the term (j tan $\omega T/2$) in $H_2(j\omega)$ varies between these same limits of j0 and j∞ as ω varies between 0 and π/T. Therefore the complete frequency-response characteristics of the analogue filter are compressed into the frequency range $0 < \omega < \pi/T$ in

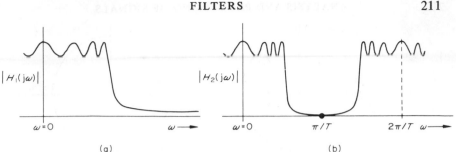

Figure 9.26   *Response magnitude characteristics of (a) seventh order Chebychev analogue filter, and (b) an equivalent digital filter based upon the transformation*
$$s \rightarrow (z - 1)/(z + 1)$$

the corresponding digital filter. This effect is illustrated by figure 9.26, which shows the response magnitude characteristics of a Chebychev low-pass analogue filter and of its digital equivalent based upon the above transformation. The response of the analogue filter falls to zero as $\omega \rightarrow \infty$, whereas that of the digital filter reaches zero at $\omega = \pi/T$, and then repeats. The compression effect becomes much more marked as the frequency approaches π/T, and has the effect of distorting the frequency scale.

The substitution $s \rightarrow (z - 1)/(z + 1)$ is one example (and a very useful one) of a number of so-called 'bilinear transformations'[27] which, in effect, compress the complete imaginary axis in the s-plane into a single revolution of the unit circle in the z-plane. As we have shown above, this particular substitution is equivalent to replacing $j\omega$ by $j \tan \omega T/2$, or ω by $\tan \omega T/2$, in a frequency-response expression. Thus, if for example, we use this technique to derive a family of digital low-pass Butterworth filters,[26] their response magnitude characteristics will take the form

$$| H(j\omega) |^2 = \frac{k}{1 + \left[\tan \dfrac{\omega T}{2} \Big/ \tan \dfrac{\omega_0 T}{2} \right]^{2n}}$$

where k is a constant.

So far we have not considered the z-plane pole–zero configurations of filters based upon such a transformation. However, when we substitute $F(z)$ for s in $H_1(s)$, we obtain a new function $H_2(z)$ which may be arranged in the form of a numerator and denominator polynomial in z: the factors of these polynomials would indicate the poles and zeros. Actually, the polynomials themselves allow us to define the recurrence formula of the filter, so that it is not strictly necessary to find the pole–zero locations. It is however worth noting that the above type of transformation converts an nth order analogue Butterworth or Chebychev low-pass filter, which has n s-plane poles, into a digital filter with n z-plane poles and n z-plane zeros (the latter all being placed at $z = -1$): this means that each output sample value is obtained from $(n + 1)$ input samples and n previous outputs in a recursive operation. Equivalent high-pass filters may be obtained by reflecting the pole–zero configurations of low-pass filters about the imaginary axis in the z-plane. A typical example of this technique is shown in figure 9.27.

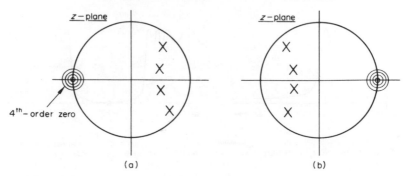

Figure 9.27   *(a) Pole-zero configuration of a fourth order digital filter with Butterworth low-pass characteristics, and (b) of its high-pass equivalent*

As we have already noted, the bilinear transformation produces a distortion of the frequency scale, which is much compressed as we approach $\omega = \pi/T$. Note, however, that Butterworth filters are designed for maximally-flat, and Chebychev for equi-ripple, passband *amplitude* characteristics. These desirable properties are not destroyed by compression of the *frequency* scale. We must not, of course, assume that only Butterworth and Chebychev characteristics are ever required; but they are, in practice, so widely used in analogue systems that any method which effectively transforms them into digital equivalents is bound to be attractive.

9.4.6 Further aspects of filter implementation

The practical implementation of a digital filter is often affected by the problem of finite 'wordlength'. Because a numerical value (representing, for example, a signal sample value or a filter multiplier coefficient) must be represented in terms of a finite number of binary digits, or bits, it is always subject to some inaccuracy. This can give rise to various difficulties, discussed briefly below. A fuller treatment can be found in almost any specialist text on digital filters (for example, that by Terrell).[46]

Perhaps the most obvious problem is that sample values can only be represented with finite accuracy. Suppose that an input signal is sampled by an 8-bit analogue-to-digital converter (ADC), prior to processing by a digital filter. Since a binary code with 8-bit wordlength can only represent 2^8, or 256, distinct signal levels, individual sample values may be in error by up to about ±0·2% of full scale. The situation becomes relatively worse if the full range of the ADC is not used. Such 'quantisation' errors in the digital representation of a signal may generally be thought of as wideband noise superimposed upon it, and may, of course, be reduced by using a longer wordlength. As digital technology advances and becomes cheaper, there is indeed a trend towards the use of 12, 16, 32 or even 64-bit words.

Digital filter multiplier coefficients may also only be represented with finite accuracy. For example, in section 9.4.3 we have considered a bandpass filter with the recurrence formula

$$y(n) = 1 \cdot 6650\, y(n-1) - 0 \cdot 9025\, y(n-2) + x(n) - x(n-2)$$

Errors in the coefficients will obviously cause the frequency response of the filter to depart to some extent from that desired. Even more seriously, a recursive filter may become unstable if its recursive multipliers are inaccurately specified, because its z-plane poles may, in effect, move outside the unit circle. The risk of this happening clearly depends on how close the poles are to the unit circle in the first place, but, as a rough guide, selective filters may require their recursive multipliers specified to 4 or 5 decimal figure accuracy — say one part in 50 000. It is worth noting that a 16-bit binary code gives a resolution of one part in about 65 000. One advantage of the recursive filters described in section 9.4.4.5 is that their multipliers, being small integers, may be specified with complete accuracy by just a few bits.

Further difficulties may arise in performing the multiplication and addition/subtraction operations required by a digital filter. Put very briefly, it is often necessary to truncate or round off the results of such computations in order to contain them within the available wordlength. This gives rise to errors, and also often leads to undesirable oscillations ('limit cycle' or 'overflow' oscillations) at the filter output. To a marked extent such effects depend upon the type of arithmetic used ('fixed-point' or 'floating-point'), and may be reduced or even eliminated by careful ordering of the various arithmetic operations required by the filter.

Although not related directly to the problem of wordlength, a final comment about filter multiplications is in order here. We have noted several times in this chapter that these normally limit the speed of operation, especially when the coefficients are decimal numbers. However, in the last few years it has been increasingly realised that the multiplications should really be termed 'pseudo-multiplications' since one of the inputs — the filter coefficient — is fixed. This allows the use of look-up tables and what is termed 'distributed arithmetic', rather than true multiplication operations, and can give valuable increases in filter speed. A summary of the technique, together with results for a variety of microprocessor systems, is given in a recent paper.[47]

Problems

1. Sketch the frequency response magnitude characteristics of the two analogue filters whose pole-zero configurations are shown in the figure. On the basis of these curves, what can you infer about the form of their impulse responses?

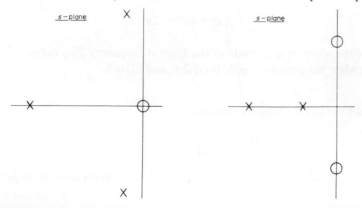

2. An analogue filter has the impulse response shown in the accompanying figure. Using the graphical interpretation of the convolution integral, infer the magnitude of its response to

  (i) a zero frequency input signal, and
  (ii) a very high frequency input signal.

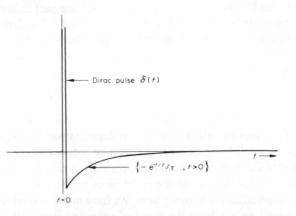

3. A Butterworth low-pass analogue filter has a nominal cut-off frequency of ω_0 radians/second. Estimate the minimum order of filter which will achieve:

  (a) a reduction in relative response magnitude to $0 \cdot 2$ at a frequency $2\omega_0$ radians/second, and
  (b) a reduction in relative response magnitude to $0 \cdot 2$ at a frequency $3\omega_0$ radians/second.

Sketch the s-plane pole configuration in each case.

4. A low-pass analogue filter with Chebychev characteristics has a nominal cut-off frequency of ω_0 radians/second. Given that the Chebychev polynomial of third order ($n = 3$) takes the form

$$C_n(x) = 4x^3 - 3x,$$

estimate the response magnitude of the filter at frequency $2\omega_0$ radians/second, when the passband ripple is (i) $0 \cdot 1$, and (ii) $0 \cdot 3$.

5. A digital filter has the transfer function

$$H(z) = \frac{z}{(z + 0 \cdot 9)}.$$

Sketch its z-plane pole-zero configuration, and the magnitude of its frequency-response characteristic in the range

$$-\frac{2\pi}{T} < \omega < \frac{2\pi}{T}.$$

Derive the recursive recurrence formula relating input and output sample values, and hence sketch the filter's impulse response.

6. Write down the non-recursive and recursive recurrence formulae for a moving-average low-pass digital filter having 12 equal impulse response coefficients.

   (i) Sketch its step response.
   (ii) Sketch its frequency-response magnitude characteristic in the range $0 < f < 1/2T$ Hz.

   If you have access to a suitable digital computer, evaluate the frequency response for (say) 100 frequency values over this range, and compare with your sketch. Convolute the impulse response of the filter with a sampled sinusoid of frequency $1/6T$ Hz, switched on at $t = 0$, and show that the filter displays a start-up transient equal in duration to its impulse response, before settling to zero steady-state output.

7. Calculate the coefficients of a 9-term Hanning window. Use a digital computer to evaluate the frequency response (say for 100 values of frequency in the range $0 < \omega < \pi/T$ radians per second) of a 9-term moving-average low-pass filter with equal coefficients, and of the same filter after application of the Hanning window. Compare the initial cut-off slope and sidelobe performance of the two designs.

8. A low-pass digital filter has the transfer function

$$H(z) = \frac{(1 - z^4)^2}{z^6(1 - z)^2}$$

   Sketch its pole–zero configuration, frequency response and impulse response. What is the computational advantage of realising the filter recursively rather than non-recursively?

9. The two sampled-data signals illustrated in the figure are applied to the digital filter of problem 8. Using the recurrence formula already derived, evaluate the form of the output in each case, and comment.

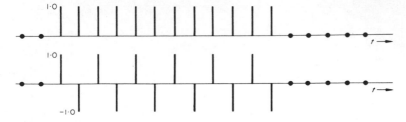

10. Figure 9.24(b) in the main text shows the pole–zero configuration for each of a bank of bandpass filters having the same peak gain but different centre frequencies. Derive the time-domain recurrence formulae for two such filters, with centre frequencies of 9 and 10 Hz, using a 100 Hz sampling rate. Place the conjugate pole-pair on a circle of radius 0·97.

Find the approximate 3 dB bandwidth of the filters, that is, the frequency range over which the response is within 3 decibels (equivalent to a factor of $1/\sqrt{2}$) of its peak value. Note that this is most simply done by considering changes in the length of pole vector P_1, for frequencies very close to the centre frequency.

11. A first-order low-pass analogue filter composed of one resistor R and one capacitor C has the impulse response

$$I(t) = \frac{1}{\tau} e^{-t/\tau}, \quad \text{for } t > 0$$

where $\tau = CR$ is the time-constant of the filter. Find the frequency response of the filter by Fourier transformation, and sketch its magnitude characteristic in the range $0 < \omega < 5/\tau$ radians per second.

It is decided to design an equivalent impulse-invariant digital filter, with a sampling period $T = \pi\tau/3$. Find the first three values of its impulse response (at $t = 0$, T, and $2T$). Would this choice of sampling period give serious distortion due to aliasing?

12. A low-pass analogue filter has the transfer function

$$G(s) = \frac{2}{(s + 1)(s + 2)}$$

Use the bilinear transformation $s \rightarrow (z - 1)/(z + 1)$ to convert the filter into a digital equivalent, and find the latter's time-domain recurrence formula. Sketch the frequency-response magnitude characteristic of both filters over suitable frequency ranges.

10

Signal Recovery, Detection and Prediction

10.1 Introduction

In this chapter, we discuss a number of problems having a strong statistical content. From a practical point of view, the most important is that of extracting a signal from random noise: it arises to some extent in almost all communication, measurement and recording situations. Another interesting problem is that of predicting the future of a random signal as accurately as possible. Not surprisingly, we shall find the earlier discussions of random signals (especially section 5.4) and linear systems (especially section 7.4) useful; and to these will be added some new perspectives in the processing of random signals and noise.

Terms such as recovery, detection, and prediction, when applied to signals and data, are not entirely standard. It is therefore important to be clear how they will be used in this chapter:

(a) By *recovery*, we shall mean the extraction of a signal from noise, when the signal waveshape is not known and must be preserved. This problem arises very generally — for example, in the reception of television video signals.
(b) The term *detection* will be used when a known signal waveshape of finite duration is contaminated by noise, and we wish only to find out when (or if) it occurs. This type of problem arises in, for example, pulse radar systems. Note that *detection* is quite widely used in the literature to include signal *recovery* as defined above.
(c) *Prediction* will mean forecasting the future of a random signal on the basis of its past history and its known statistical properties. An example of this problem is predicting the future position of an aircraft, when the pilot of that aircraft is trying to take evasive action. Because movements of the aircraft are limited by its dynamics, such action should be, to some extent, predictable.

When a signal, mixed with unwanted random noise, is to be recovered or detected by a linear processor, we may fortunately use the principle of super-position. In other words, the effects of the processor on the signal and noise components may be considered separately. This is a great advantage. It assumes, of course, that the noise is additive; this is a valid assumption in many practical situations. Other types of noise contamination (such as multiplicative) do sometimes occur, but are much harder to analyse and will not be discussed here.

Whether a random waveform is termed 'signal' or 'noise' depends on one's point of view: if wanted, it is a signal; if unwanted, it is noise. Therefore much of the work we have done on random signals in chapters 5, 6 and 7 is just as applicable to the description and processing of random noise.

10.2 Signal recovery

10.2.1 Signals in wideband noise

Suppose a signal, occupying a well-defined frequency band, becomes contaminated with noise of much wider bandwidth. This often happens in, for example, radio, radar and television transmission systems. A typical situation of this type is illustrated in figure 10.1. The most obvious filter to use is one which transmits all signal components equally, but rejects all others — as indicated in the figure. If distortion of the signal waveshape is to be avoided, the filter's phase response should be as linear as possible. As we have noted in section 9.2.2, no filter can display these idealised characteristics, so in practice we must settle for a finite cut-off slope in the region of $\omega = \omega_c$.

Before developing this theme further, we should be clear that this simple and widely used approach for improving the signal-to-noise (S:N) ratio is not necessarily *optimum*, in the sense of minimising the r.m.s. error between the desired filter output (the 'pure' signal) and its actual output (filtered signal-plus-noise). It achieves the best results only when signal and noise energy lying within the band $-\omega_c < \omega < \omega_c$ are similarly distributed within it. But if, for example, the noise spectrum had a pronounced peak at some frequency ω_n within the signal's frequency range, a filter with the characteristic shown in figure 10.1 would allow

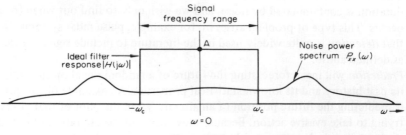

Figure 10.1

through a lot of noise energy. We must expect that a better result would be obtained using a filter with a transmission trough in the region $\omega = \omega_n$, since any resulting distortion of the signal would be more than offset by the large reduction of noise. To convert such qualitative arguments into quantitative ones involves the theory of optimum filters, introduced in section 10.2.4.

Returning to the simpler approach illustrated by figure 10.1, the improvement in S:N ratio obtained will obviously depend greatly on the frequency range, or bandwidth, of the noise: if it is the same as that of the signal $(-\omega_c < \omega < \omega_c)$ there will clearly be no improvement at all. In the general case, let us denote the power spectrum of the noise by $P_{xx}(\omega)$. The total noise power (that is, mean square value) prior to filtering is then

$$\frac{1}{2\pi} \int_{-\infty}^{\infty} P_{xx}(\omega) \, d\omega$$

which is reduced at the output of the filter to

$$\frac{1}{2\pi} \int_{-\infty}^{\infty} P_{xx}(\omega) \, | \, H(j\omega) \, |^2 \, d\omega$$

where $H(j\omega)$ is the frequency response of the filter. Assuming the idealised filter characteristic of figure 10.1, this latter expression may be written as

$$\frac{A^2}{2\pi} \int_{-\omega_c}^{\omega_c} P_{xx}(\omega) \, d\omega$$

so that the ratio R of noise power at input and output of the filter is

$$R = \frac{\displaystyle\int_{-\infty}^{\infty} P_{xx}(\omega) \, d\omega}{A^2 \displaystyle\int_{-\omega_c}^{\omega_c} P_{xx}(\omega) \, d\omega}$$

The ratio of noise amplitude, measured in terms of the square root of its mean square (r.m.s.) value, will therefore be $\sqrt{R}$. The signal amplitude at the filter output will be A times that at the input. If we express the improvement in signal-to-noise (S:N) ratio due to the filter in terms of peak signal and r.m.s. noise, we have

$$\text{S : N ratio improvement} = \frac{\text{peak signal out}}{\text{peak signal in}} \times \frac{\text{r.m.s. noise in}}{\text{r.m.s noise out}}$$

$$= A\sqrt{R}$$

$$= \left[\frac{\displaystyle\int_{-\infty}^{\infty} P_{xx}(\omega) \, d\omega}{\displaystyle\int_{-\omega_c}^{\omega_c} P_{xx}(\omega) \, d\omega} \right]^{1/2}$$

This result is greatly simplified when the noise has a flat power spectrum (that is, is 'white') between limits of, say, $-\omega_n < \omega < \omega_n$, with negligible noise power outside this frequency range. $P_{xx}(\omega)$ is then a constant over the frequency range $-\omega_n < \omega < \omega_n$, giving

$$S : N \text{ ratio improvement} = \left[\frac{\omega_n}{\omega_c} \right]^{1/2}$$

Figure 10.2 illustrates a typical case, in which the noise at the input occupies a frequency range about nine times greater than that of the signal: its r.m.s. value at the output is therefore reduced relative to the signal by a factor of about three. Note also that the noise is now different in form, since it contains only components lying within the filter passband.

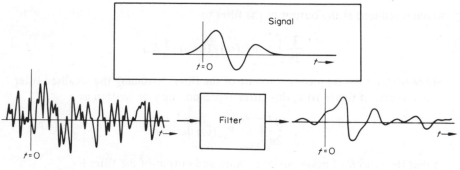

Figure 10.2   *Recovery of a signal from wideband noise. The filter transmits all signal components equally, but rejects noise lying outside the signal's frequency range. Shown inset is the pure signal waveform*

10.2.2 Signals in narrowband noise

So far we have considered a relatively narrowband signal mixed with wideband noise. At the other extreme, a wideband signal may be contaminated with narrowband noise. A good example is the pick-up of mains-frequency (50 Hz) interference during signal recording. Strictly, it may be argued that such mains 'hum' is not noise, because its frequency is well defined and its waveform is close to sinusoidal. However, chance fluctuations in its amplitude during recording can give it many of the properties of narrowband random noise. In such a case the obvious filter to use is one with a notch characteristic centred at mains frequency. A good design will transmit all signal frequency components, except those very close to the notch frequency, with constant gain and negligible phase distortion.

The application of a simple digital notch filter to electrocardiogram (ECG) processing is illustrated in figure 10.3. An ECG waveform, representing electrical

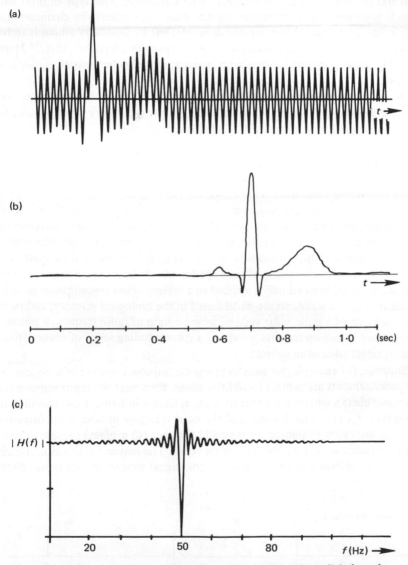

Figure 10.3 *Recovery of a signal from narrowband noise, using a digital notch filter: (a) the contaminated signal, (b) the recovered signal, and (c) the filter characteristic. Waveforms (a) and (b) are shown in reconstituted (analogue) form, for clarity*

activity of the heart, is shown at the top of the figure, heavily contaminated with 50 Hz interference. In the middle is shown the filter output — a relatively pure ECG waveform, with almost all the narrowband interference removed. The filter's frequency-response magnitude characteristic, with its notch centred at 50 Hz, is shown below. Since the signal energy is quite widely distributed over the range

0 to 200 Hz, the notch has little effect on its waveshape. This type of filter has already been mentioned in section 9.4.4.5. True, the ripples in its characteristic to either side of the main notch are not desirable, but its simplicity allows it to be programmed for real-time operation on an inexpensive microcomputer.[43] If needed, digital (or analogue) notch filters with better amplitude characteristics may certainly be designed.[40]

We should note, once again, that the above approach is not optimum in any rigorous sense. Nevertheless, when the noise bandwidth is very much smaller than that of the signal, it is normally very adequate in practice.

10.2.3 Signal averaging

When recovering a signal from noise, we need to take advantage of all its known characteristics. So far, in considering a signal in wideband or narrowband noise, our only assumptions have been about bandwidth. If, however, a signal waveform of finite duration, obscured by noise, is known to repeat itself at particular instants of time, a further technique known as 'signal averaging' may be used to clarify it. Such a situation arises quite commonly in experiments of the stimulus-response type, where a series of brief stimuli is applied to a system under investigation in order to excite it. Such experiments are widely used in the biological sciences; and parallel the operation of a pulse radar system, where a train of radio frequency pulses, transmitted at known instants, produces a corresponding series of echoes from a distant target such as an aircraft.

Suppose, for example, we wish to assess the impulse response of a system, but our measurements are subject to additive noise. We repeat the input impulse many times and elicit a whole series of responses, as shown in figure 10.4. Provided the instants $t_1, t_2, t_3 \ldots$ are known, and the noise is largely or wholly uncorrelated between successive responses, we may add together (or average) successive versions of the response to derive a more accurate result. The reason for this is that each response is composed of two parts — the true signal waveform, and noise. When we

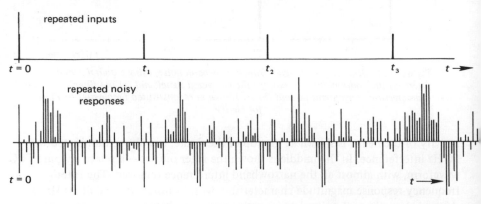

Figure 10.4

add a number of them together the signal portion is truly additive, whereas the noise, being uncorrelated, tends to average out to a constant level which reflects its mean value.

If the noise is completely uncorrelated between successive versions of the response, the improvement in signal-to-noise ratio due to averaging may easily be estimated. As already mentioned in section 5.5.2, the variance of the sum of a number of statistically independent (that is, uncorrelated) random signals is equal to the sum of their individual variances. Therefore if we sum n versions of the noisy response, we get a signal increased in size by a factor n, and a noise variance increased by the same factor; the standard deviation of the noise therefore increases by $\sqrt{n}$. The ratio of signal amplitude to noise standard deviation is therefore improved by $\sqrt{n}$. Figure 10.5 shows the typical result of averaging 8, 32 and 128 responses of the type already illustrated in figure 10.24. If the noise contains significant energy at low enough frequencies, it will tend to be correlated between successive input impulses, and the improvement in signal-to-noise ratio will be less than $\sqrt{n}$: it will also be more difficult to predict theoretically. A somewhat different problem arises if the noise contains strictly periodic components (such as at mains-supply frequency): in such a case the noise will clearly be correlated between successive presentations of the input if the latter is also strictly periodic. A convenient way of avoiding this difficulty is to apply the input impulse at irregular instants.

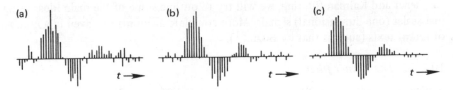

Figure 10.5   *Typical results of averaging (a) 8, (b) 32, and (c) 128 noisy responses of the type shown in figure 10.4*

Finally, we should note that filtering of the type already discussed in sections 10.2.1 and 10.2.2 may be used *in addition* to signal averaging, to further enhance the S:N ratio — assuming, of course, that we have information about signal and noise bandwidths, as well as about times of occurrence of the signal. In the example illustrated in figures 10.4 and 10.5, it is clear that the noise is of much wider bandwidth than the signal, so that such additional filtering would produce benefits.

10.2.4 *Optimum signal estimation*

10.2.4.1 *Introduction*

Our discussion of analogue and digital filters in Chapter 9 concentrated very largely on frequency-domain properties, on the assumption that the majority of filtering problems can be specified in terms of desired bandpass or bandstop characteristics.

This preoccupation with frequency responses and frequency spectra is largely historical, since the development of filter theory (and particularly analogue filter theory) has been closely tied up with radio communication, and its need to distinguish between signals on the basis of their different spectral bands. As we have seen in sections 10.2.1 and 10.2.2, this approach can also be valuable for recovering signals from noise.

However, the widespread availability of digital hardware, computers and microprocessors has opened up new possibilities for signal recovery over the past fifteen or twenty years. Increasing emphasis has been placed on optimum filtering in the time-domain, in the sense of minimising the mean square error between the desired filter output (the 'pure' signal) and its actual output (filtered signal-plus-noise). The choice of mean square error as the criterion is both sensible and convenient: sensible, because it takes equal account of positive and negative instantaneous errors — unlike the mean error, which could well average out to zero; and convenient because, as we shall see, it ties in neatly with the theory of linear processors and correlation functions. As already noted, such optimisation is not, in general, achieved by frequency-domain methods.

The design of optimum filters for processing signals in noise is often referred to as an *estimation* problem. That is to say, given a signal corrupted by noise, how do we specify the filter which achieves the best estimate of the signal in a 'least-squares' sense? This is rather a complicated matter, normally discussed in terms of signal vectors (mentioned in section 2.3.1) and matrix theory. In this brief introduction to Wiener and Kalman filtering, we will try to outline some of the main ideas using just scalar (one-dimensional) signals. More complete accounts are given in a number of recent texts (such as that by Bozic[41]).

10.2.4.2 *The Wiener filter*

A Wiener filter is essentially a non-recursive digital filter, designed to produce an optimum output in the least-squares sense. A Kalman filter uses recursive techniques to achieve the same objective. To illustrate the Wiener filter, consider the sampled signal-plus-noise waveform of figure 10.6. A sample value such as $x(n)$ consists of

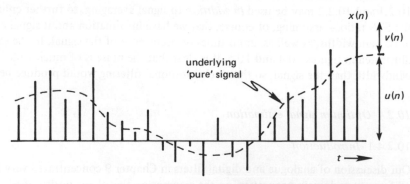

Figure 10.6   *A noisy signal waveform (x), consisting of a pure signal (u) corrupted by additive noise (v)*

two parts: a 'pure' signal component $u(n)$ plus an additive random noise component $v(n)$. Note that $v(n)$ may in general be either positive or negative. We now wish to design a filter to give us the best estimate, on average, of the 'pure' signal. Suppose our non-recursive (Wiener) filter has impulse response terms $a_0, a_1, a_2, \ldots, a_k$. Its output is given by

$$y(n) = a_0 x(n) + a_1 x(n-1) + \ldots + a_k x(n-k) = \sum_{i=0}^{k} a_i x(n-i)$$

This output is, of course, to be the estimate of the true signal value $u(n)$. Hence the error is given by

$$e(n) = u(n) - y(n) = u(n) - \sum_{i=0}^{k} a_i x(n-i)$$

and we must now adjust the filter coefficients to minimise the average, or expected, value of the square of this error, taken over a large number of output sample values. Denoting an expected value by E, we wish to minimise

$$P = E[e(n)^2] = E\left[u(n) - \sum_{i=0}^{k} a_i x(n-i) \right]^2$$

This may be done by differentiating with respect to each of the coefficients, and equating the result to zero. Thus, for the jth coefficient, a_j, we have

$$\frac{\partial P}{\partial a_j} = -2E\left[u(n) - \sum_{i=0}^{k} a_i x(n-i) \right] . x(n-j) = 0$$

therefore

$$E\left[\sum_{i=0}^{k} a_i . x(n-i) . x(n-j) \right] = E\left[u(n) . x(n-j)\right]$$

or

$$\sum_{i=0}^{k} a_i . E[x(n-i) . x(n-j)] = E[u(n) . x(n-j)]$$

Now the expected, or average, value of the product $x(n-i) . x(n-j)$ is just the input autocorrelation function (ACF) relevant to a time-shift of $(i-j)$ sampling periods. Let us denote this by $r_{11}(i, j)$. Similarly, the expected value of the product $u(n) . x(n-j)$ is equal to the crosscorrelation function (CCF) of input and desired output (the pure signal), for a time-shift of j sampling periods. Denoting this by $r_{12}(j)$, we may write

$$\sum_{i=0}^{k} a_i . r_{11}(i, j) = r_{12}(j) \quad \text{for } j = 0, 1, 2, \ldots k$$

This important result is known as the Wiener-Hopf equation. It shows that we may evaluate the optimum filter coefficients if we know the ACF of the noisy input signal, and can specify the CCF between this input and the desired output for positive values of time-shift.

As it stands, the Wiener–Hopf equation is hard to visualise. Let us therefore take a simple illustrative example: a constant signal value $u(n) = 1 \cdot 0$, corrupted by additive white noise with zero mean and unit variance. A typical portion of the noisy input to the estimating filter is shown in figure 10.7(a). Suppose we decide

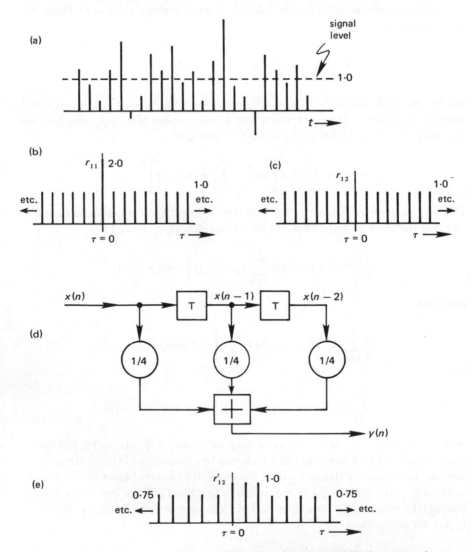

Figure 10.7   *Illustration of Wiener filtering: (a) the noisy signal to be estimated, (b) the input ACF, (c) the desired input–output CCF, (d) the optimum 3-term Wiener filter, and (e) the actual input–output CCF produced by the filter*

to use just a three-term Wiener filter to estimate the true signal level of $1 \cdot 0$. Let us now find its coefficients a_0, a_1, and a_2. First we should bear in mind that an ACF is always symmetrical in form (see section 5.4.2), so that $r_{11}(i, j)$ equals $r_{11}(j, i)$. Further, for simplicity, let us write $r_{11}(1)$ to denote $r_{11}(i, j) = r_{11}(j, i)$ when $i - j = 1$, etc. The Wiener–Hopf equation then yields

$$r_{11}(0) \cdot a_0 + r_{11}(1) \cdot a_1 + r_{11}(2) \cdot a_2 = r_{12}(0) \quad \text{for } j = 0$$

$$r_{11}(1) \cdot a_0 + r_{11}(0) \cdot a_1 + r_{11}(1) \cdot a_2 = r_{12}(1) \quad \text{for } j = 1$$

$$r_{11}(2) \cdot a_0 + r_{11}(1) \cdot a_1 + r_{11}(0) \cdot a_2 = r_{12}(2) \quad \text{for } j = 2$$

We next need the values of the two correlation functions r_{11} and r_{12} for the relevant values of time-shift. The ACF of the noisy input signal is shown in figure 10.7(b). The constant signal value $u(n) = 1 \cdot 0$ produces a constant ACF value, also of $1 \cdot 0$; the noise gives an additional contribution at zero shift ($\tau = 0$) equal to its mean square value (equal to its variance of $1 \cdot 0$, since its mean level is zero) – but no contribution elsewhere because it is white (successive values independent). The CCF between input and desired output is shown in part (c) of the figure: since the desired output is just the steady signal value of $1 \cdot 0$, and this is uncorrelated with the noise part of the input, the CCF is also a series of $+1 \cdot 0$ values. Hence we see that

$$r_{11}(0) = 2; \ r_{11}(1) = r_{11}(2) = 1 \cdot 0; \ r_{12}(0) = r_{12}(1) = r_{12}(2) = 1 \cdot 0$$

When these values are substituted into the above equations, it is easy to show that

$$a_0 = a_1 = a_2 = 1/4$$

Therefore in this case the optimum three-term non-recursive estimator, or Wiener filter, is as shown in figure 10.7(d).

Before discussing this result, it must be admitted that the simple example we have chosen is somewhat artificial. For if the signal value $u(n)$ is constant and known in advance – as has been assumed in working out the autocorrelation function r_{11} – we hardly need an optimum filter to estimate it. Note, however, that the same filter would apply to a more interesting situation: a signal $u(n)$ which varied very slowly, and was therefore effectively constant in the short term; but which could, in the long term, fall anywhere in a range having unit mean square value. In this case the first few values of the correlation functions r_{11} and r_{12} would be similar to those shown in figure 10.7(b) and (c). So therefore would be the Wiener filter.

It is not surprising that, in the case we have investigated, the optimum filter turns out to have three *equal* coefficients. After all, the signal is assumed constant and the noise is stationary, so any input sample $x(n), x(n - 1), x(n - 2), \ldots$, is just as good a source of information about the 'pure' signal as any other. If however, as is often the case, the signal varies quite rapidly, then its present and *immediate* past history is most valuable for estimating its true *present* value. Hence the coefficients a_j of the Wiener filter would tend to fall off in absolute value with increasing j. There is, indeed, little to be gained by choosing a large number of coefficients in such a case.

Although the filter coefficients of figure 10.7(d) might be expected to be equal, their actual value is, at first sight, surprising. If there was no noise, the signal value would clearly be *exactly* estimated by three coefficients of value $1/3$ — a true moving-average filter (see section 9.4.4.2). The reason why smaller coefficients give a better performance when noise is present may be summarised as follows. Suppose the signal value was very much smaller than the noise level (for example, signal = 0.01, noise variance = 1.0). Even though the filter *tends* to average the noise out to zero, this desirable effect is not very marked with only a three-term filter, and a lot of noise would reach the filter output, swamping the signal. The instantaneous error would therefore generally be large. Our optimum filter recognises this fact and by reducing the coefficients, also reduces the level of output noise. In doing this, it also introduces an error in the 'pure' signal output, but the trade-off between these two effects is optimum. As the $S:N$ ratio worsens, the filter coefficients become smaller and smaller. We therefore see that the design of such a filter depends not only on the spectral distribution of signal and noise, but also on their relative levels. This is an important and surprising feature of such optimal filtering. We should also note that because of the error introduced by the filter in the pure signal output, the estimate is 'biased'. In other words its average, or expected, output value is not that of the signal. The best *unbiased* estimator is, in fact, the moving-average filter (with coefficients equal to $1/3$); but its mean square error is greater than that of the Wiener filter.

We saw earlier that the Wiener—Hopf equation requires the CCF between the filter input and the desired output to be specified. It is interesting to compare this with the CCF between input and *actual* output produced by our three-term filter. In our discussion of system identification by input—output crosscorrelation in section 7.4 it was shown that the input—output CCF across any linear processor is equal to the ACF of the input, convoluted with the processor's impulse response. Applied to the present example, this gives the result shown in figure 10.7(e). We see that the actual CCF $r'_{12}(j)$ is equal to the desired one $r_{12}(j)$ for $j = 0$, 1 and 2, but is different elsewhere. This illustrates the general result that a Wiener filter with k impulse response coefficients produces k correct terms in the CCF, at positive values of time-shift. Errors in the remaining terms represent, of course, the imperfect action of the filter.

In spite of its elegance, the Wiener filtering technique involves several problems. Firstly, the ACF of the noisy input signal must be known in advance; this is tantamount to knowing the power spectral distributions of both signal and noise. Secondly, the number of filter coefficients must be specified, and if subsequently changed all the coefficients must be recalculated. Of course, the more coefficients are used, the more accurate the estimator, but the greater the amount of computation involved. The third problem is that a filter with k coefficients requires the solution of k simultaneous equations. We have only used three, but if a much larger number is specified their calculation using the Wiener—Hopf equation will require a digital computer.

10.2.4.3 *The Kalman filter*

A Kalman filter uses recursive techniques to achieve optimum signal estimation. Once again, we will limit ourselves to illustrating some of the main features of this approach by a simple example. Whereas the majority of texts on Kalman filtering use matrix theory and signal vectors, we will consider just scalar signals. On the left-hand side of figure 10.8, within the dashed lines, is a mathematical model of the signal we wish to estimate. It covers the generation of the signal and its subsequent measurement, and is based upon whatever information is available to us. The model has three essential aspects: a zero-mean white noise source; a linear processor, taking the form of a recursive digital filter, which gives the noise spectral properties similar to those of the actual signal; and additive white measurement noise, again with zero-mean value, representing random errors during measurement or observation. As with the Wiener filter, the only quantity available to us is the noisy signal $x(n)$, and we process this with a Kalman filter to produce an optimum estimate $y(n)$ of the 'pure' signal $u(n)$.

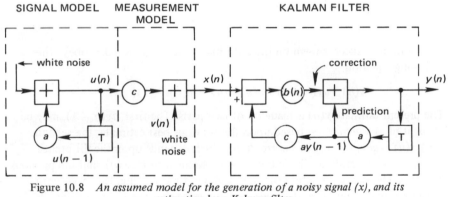

Figure 10.8   *An assumed model for the generation of a noisy signal (x), and its estimation by a Kalman filter*

Our example for a signal model is about the simplest possible: a single z-plane pole at $z = a$ is used to modify the spectral properties of the white noise. Figure 10.9 shows typical portions of signals generated by this model, with $a = 0.5$, 0.98 and -0.98. When $a = 0.5$, the low-pass filtering effect of the pole is already apparent, giving the 'signal' a predominantly low-frequency appearance. This is much more marked with $a = 0.98$. When a is negative, the high-frequency content of the noise is emphasised. Setting it to -0.98 produces a very clear high-pass action, with adjacent sample values almost always alternating in sign. The figure therefore illustrates the type of random signal which can be appropriately represented using a first-order recursive filter; but it must be stressed that more complicated linear processors will often be required to model the signal with reasonable accuracy. The 'pure' signal $u(n)$ is then modified by a measurement parameter c (which may be thought of as a gain or amplification factor) and by additive white noise, to produce the 'noisy' signal $x(n)$.

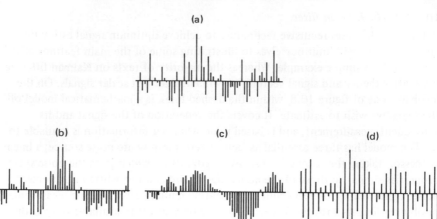

Figure 10.9   *Typical signal portions (u) generated by the model on the left-hand side of figure 10.8: (a) shows the white noise input to the model; (b), (c) and (d) show model outputs for a = 0·5, 0·98, and −0·98 respectively*

The Kalman filter, shown on the right-hand side of figure 10.8, obeys the following equation

$$y(n) = ay(n-1) + b(n)[x(n) - acy(n-1)]$$

The present estimate $y(n)$ is made up of two parts. The first, $ay(n-1)$, may be thought of as a *prediction* based upon the previous best estimate; the second, $b(n)[x(n) - acy(n-1)]$ as a *correction*, which depends upon the difference between the currently available input to the Kalman filter, $x(n)$, and the previous estimate. The gain parameter $b(n)$ is, in general, time-varying. However, with stationary signal and noise statistics, and assuming that the Kalman filter has reached its steady-state operation (that is, any start-up transient is completed), $b(n)$ itself reaches a steady value b related to the parameters a and c by the equation[41]

$$b^2(a^2 c\sigma_V^2) + b(c^2\sigma_W^2 + \sigma_V^2[1-a^2]) - c\sigma_W^2 = 0$$

where σ_W^2 and σ_V^2 are the variances of the white noise sources generating the signal and measurement noise respectively. It may be shown that this filter is an optimum recursive estimator of the signal $u(n)$.

Let us take a simple example, and evaluate some parameters. Suppose $c = 1$; and $a = 0·98$, giving a signal $u(n)$ which tends to vary quite slowly (see figure 10.9(c)). Let us also arrange that both the pure signal $u(n)$ and the measurement noise $v(n)$ have unit variance (this will allow a comparison with the Wiener filter example of figure 10.7). We must now evaluate the variance σ_W^2 of the white noise source generating $u(n)$. Note that the impulse response of the single-pole recursive filter in our model has successive terms

$$1, a, a^2, a^3, a^4 \ldots \text{(see also section 9.4.2 and figure 9.14)}$$

Hence the filter has an exact non-recursive equivalent with transversal multipliers also equal to $1, a, a^2, a^3, a^4, \ldots$, etc. Consider this filter fed with white noise of variance $\sigma_W{}^2$: from each tapping point will come independent noise, to be weighted in variance by the square of one of the above coefficients, and then summed to produce $u(n)$. When we combine a number of independent, zero-mean, random processes, their variances are additive (see also section 5.5.2). Therefore the variance of $u(n)$ will be

$$\sigma_W{}^2 . 1^2 + \sigma_W{}^2 . a^2 + \sigma_W{}^2 . a^4 + \sigma_W{}^2 . u^6 + \sigma_W{}^2 . u^8 + \ldots = \frac{\sigma_W{}^2}{(1 - a^2)}$$

and, since this is to be unity, $\sigma_W{}^2$ must be equal to $(1 - a^2)$. Our chosen parameters are therefore

$$a = 0.98, c = 1, \sigma_V^2 = 1, \sigma_W^2 = (1 - a^2) = 0.0396$$

which give $b = 0.1660$, and the Kalman filter recurrence formula

$$y(n) = 0.98\, y(n - 1) + 0.1660[x(n) - 0.98\, y(n - 1)]$$

This may also be written as

$$y(n) = 0.8173\, y(n - 1) + 0.1660\, x(n)$$

As an alternative to figure 10.8, the Kalman filter may therefore be viewed as an elementary recursive filter with low-pass properties. This offers further insights into the way it operates. We see that its single z-plane pole is at $z = 0.8173$, rather than at $z = 0.98$ as in the signal model, and that input samples are weighted by the coefficient 0.166. How, in qualitative terms, can these values be explained? Firstly note that the filter's transfer function may be written as

$$G(z) = \frac{Y(z)}{X(z)} = \frac{0.166}{(1 - 0.8173\, z^{-1})}$$

and hence its maximum gain, which occurs at $\omega = 0$ (equivalent to $z = 1$) is equal to

$$\frac{0.166}{1 - 0.8173} = 0.909$$

This may seem surprising, because we might expect a filter designed to estimate a slowly varying signal to have a gain of unity at very low frequencies. But, as we saw earlier with the Wiener filter, the gain is in fact optimum at a somewhat lower value, because the signal $x(n)$ is corrupted by noise. This helps explain the coefficient of 0.166 applied to the input. Secondly, consider the impulse response of the Kalman filter. This is of decaying exponential form, with values: $0.166, 0.136, 0.111, 0.091,$ $0.074, 0.061, 0.049 \ldots$. In effect, therefore, the filter takes significant account of many input samples in forming its current estimate. Had its pole been at $z = 0.98$, even more of the signal's past history would have been taken into account; but the present value, $x(n)$, would then have been weighted by less than 0.166, in order to preserve the filter gain of about 0.9. The pole position therefore represents a compromise: the present input is in one sense the best information we have about the current 'pure' signal, and should therefore be weighted heavily — but it is unfortunately corrupted by noise; however, the estimate may be improved by taking account of the signal's past history — but we must not go too far back, or the correlation between previous and present signal values will be lost. It is this compromise which the mathematical derivation of the Kalman filter optimises.

To summarise the differences between the Wiener and Kalman optimum filters, we see that the Wiener filter has a finite impulse response (FIR), with its number of terms specified in advance; the Kalman filter is recursive, with an infinite impulse response (IIR). And whereas no particular *form* of impulse response waveshape is assumed by the Wiener method, the Kalman filter's impulse response has a form which is directly related to the assumed dynamics of the process generating the signal — in our example, represented by the parameter a. In figure 10.7(e), we saw that the three-term Wiener filter produced three correct values of the input—output crosscorrelation function, at $\tau = 0$, $\tau = T$, and $\tau = 2T$. By convoluting the input ACF with the Kalman impulse response, we can readily find the corresponding result for the Kalman filter. Not surprisingly perhaps, it turns out that the Kalman filter does not produce any *exactly* correct CCF values, but it gets a much larger number of them *nearly* right.

This brief introduction to Wiener and Kalman estimating filters has concentrated on some of their main features, and tried to relate them to our earlier work on digital filters in chapter 9. Several important aspects — perhaps, above all, the analysis of minimum error — have been omitted for lack of space. However, the reader will find these covered in any more comprehensive text on optimum filtering (for example, Bozic,[41] Gelb[48]).

10.3 Signal Detection

10.3.1 Introduction

So far, we have considered the problem of recovering a signal waveform of unknown shape, using whatever *a priori* information is available to us: the frequency band it occupies, its times of occurrence, or its time-domain statistical properties. A rather different situation arises when we know the signal waveform (that is, we know not only its frequency band but also the amplitude and relative phases of its various components within that band), and we need to establish only *when*, or indeed *if*, it occurs.

Two examples will serve to illustrate this type of *detection* problem. The classic instance in signal processing is pulse radar, let us say for air traffic control. A train of pulses is sent out by the radar transmitter, and the receiver has to detect pulse echoes of similar shape, reflected from distant aircraft. Their waveform is known in advance and is not itself the important feature; what matters is the reliable detection of the echoes, which are often weak and badly contaminated by random noise. Aircraft must not be missed but, equally, the receiver must not interpret noise peaks as echoes when no aircraft is in fact present. Our second example is probably less familiar. Using modern instrumentation and recording methods, it is often possible to detect the electrocardiogram (ECG) of an unborn baby by attaching electrodes to the mother. This electrical signal, which accompanies the heartbeat, may be used to signify that all is well with the foetus. The signal is often weak and mixed with unwanted noise, but its waveshape is roughly known in advance and the main interest is in when, or if, it occurs. Matched filtering, as described in this section, had indeed been successfully applied to the detection of foetal ECG waveforms.

10.3.2 The matched filter

The matched filter occupies a rather central role in signal theory, because it neatly illustrates many of the important relationships between convolution correlation, and filtering (see section 7.3), and because it has found widespread practical application for signal detection in the presence of noise. For reasons which will become clear later, we will discuss the matched filter by reference to sampled-data signals and filters, although the general concepts involved apply equally well to continuous ones.

Figure 10.10 shows both a time-limited sampled-data signal and a longer waveform in which the signal occurs a number of times, mixed with additive noise. A visual inspection might suggest the onset of a signal waveform at times A, B and C, although it is difficult to be confident. We now consider the type of filter which

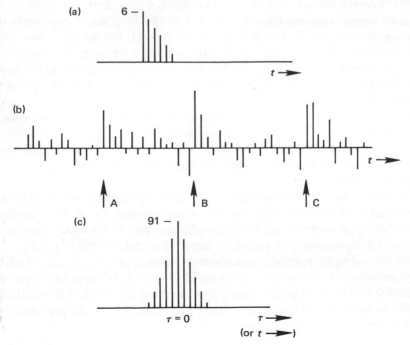

Figure 10.10   (a) A time-limited sampled-data signal, and (b) the same signal occurring several times in random noise. In (c) is shown the signal's autocorrelation function

would be most effective in enhancing the chances of correct detection of the signal.

Since the signal waveshape is assumed known, the filter need not necessarily transmit it in undistorted form. A better criterion is that the filter should give the greatest possible instantaneous output whenever the signal waveform occurs: the detection task is then reduced to searching the filtered signal-plus-noise waveform for large peaks. This is easily accomplished by an electronic circuit or computer program.

The idea of a matched filter was first introduced in section 7.3. Such a filter has an impulse response equal to a time-reversed version of the signal waveform to which it is matched: this means that its response to such a waveform is identical in form to the latter's autocorrelation function. As we have pointed out in section 5.4.2, the peak value of an autocorrelation function $r_{xx}(\tau)$ occurs when $\tau = 0$, and is proportional to the mean square value — and hence to the total energy — of the signal waveform. Thus if we process a signal-plus-noise waveform such as that of figure 10.10(b) with a matched filter, the largest peak outputs due to the signal will correspond to $r_{xx}(0)$. It may be shown theoretically that, if the noise has a constant power spectral density over the frequency range occupied by the signal, then the improvement in signal-to-noise ratio caused by a matched filter is the best possible (or, more correctly, the improvement is the best possible with any linear filter; and, in the restricted case of gaussian noise, it is the best attainable with any filter, linear or non-linear). In more detail, suppose successive samples in the signal of figure 10.10(a) have the values 6, 5, 4, 3, 2, 1. The matched filter must then have successive impulse response values 1, 2, 3, 4, 5, 6, and its output in response to the signal waveform alone will take the form shown in part (c) of the figure. Note that the peak value of 91, corresponding to $r_{xx}(0)$, will occur whenever the complete signal has just been fed into the filter. Matched filter detection therefore involves a time-delay equal to the duration of the signal waveform, but this is rarely a practical disadvantage.

At this point, it is tempting to infer that because the peak signal output value in this case is 91, whereas the peak signal input to the matched filter is only 6, the improvement in signal-to-noise (S:N) ratio must be large. But remember that the noise also passes through the filter, and is in fact substantially increased in amplitude as well.

Fortunately, the real improvement in S:N ratio is quite easily calculated if we assume that successive input noise samples are uncorrelated. This implies that the noise is 'white', at least over the full bandwidth occupied by the signal. Suppose we have input noise samples with a standard deviation σ (and variance σ^2) and a general form of signal waveform with n sample values $a_1, a_2, a_3, \ldots, a_n$, of which the largest value is $\hat{a}$. At the input to the matched filter, the ratio of peak signal to standard deviation of the noise is clearly $\hat{a}/\sigma$. Assuming the matched filter to have an impulse response consisting of sample values $a_n, \ldots, a_3, a_2, a_1$, the peak signal output from the filter is simply given by

$$r_{xx}(0) = (a_1^2 + a_2^2 + a_3^2 + \ldots + a_n^2) = \sum_{i=1}^{n} a_i^2$$

The way in which the noise is modified in passing through the filter may be visualized by recalling the graphical interpretation of the convolution integral. In this case each output noise sample is equal to the sum of n weighted input noise samples — the weighting coefficients being, of course, the terms of the impulse response. Hence each value is equal to the summation of n independent random numbers having standard deviations $a_1\sigma, a_2\sigma, a_3\sigma, \ldots, a_n\sigma$. Since variances are additive in this case, the variance of the output noise samples is therefore

$$a_1^2\sigma^2 + a_2^2\sigma^2 + a_3^2\sigma^2 + \ldots + a_n^2\sigma^2 = \sigma^2(a_1^2 + a_2^2 + a_3^2 + \ldots a_n^2)$$

The ratio of peak signal to standard deviation of noise at the filter output is

$$\frac{(a_1^2 + a_2^2 + a_3^2 + \ldots + a_n^2)}{\sigma\sqrt{(a_1^2 + a_2^2 + a_3^2 + \ldots + a_n^2)}} = \frac{\sqrt{(a_1^2 + a_2^2 + a_3^2 + \ldots + a_n^2)}}{\sigma}$$

so the filter has *improved* this ratio by a factor of

$$\frac{\sqrt{(a_1^2 + a_2^2 + a_3^2 + \ldots + a_n^2)}}{\sigma} \frac{\sigma}{\hat{a}} = \frac{\sqrt{(a_1^2 + a_2^2 + a_3^2 + \ldots + a_n^2)}}{\hat{a}}$$

$$= \frac{\left[\sum_{i=1}^{n} a_i^2\right]^{1/2}}{\hat{a}}$$

In the example we have illustrated in figure 10.10, the improvement would be $\sqrt{(91)}/6 = 1\cdot6$, or $4\cdot1$ dB. This result also shows that the benefit to be gained by use of a matched filter depends only upon the energy in the signal waveform — of which the numerator in the above expression is a measure — and upon its peak value prior to filtering. Thus the benefit would be the same for the three signals illustrated in figure 10.11 even though they have quite different forms of frequency spectra.

   As with other aspects of signal processing, corresponding theoretical results may be derived for the case of a continuous signal and an analogue matched filter: once again, the improvement in signal-to-noise ratio may be shown[3] to depend only on the total energy in the signal waveform, and not on its detailed shape. Unfortunately, however, analogue matched filters are not generally realisable, for reasons which may be summarised as follows. If a continuous signal has a spectrum $H(j\omega)$ and Laplace transform $H(s)$, then, as shown in section 7.3, its matched filter must

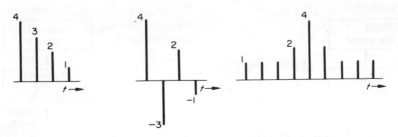

Figure 10.11 *Three signals with the same peak value and total energy.*

have a frequency response $H^*(j\omega) = H(-j\omega)$, and hence a transfer function $H(-s)$. Since a practical continuous signal has one or more poles in the left-hand half s-plane, this implies that its matched filter must have mirror-image poles in the right-hand half s-plane; such a filter would not be stable. A similar difficulty arises in the case of a sampled-data signal specified in terms of z-plane poles. However, if such a signal is specified in terms of a finite set of z-plane zeros (which implies that both signal and matched filter impulse response contain a finite number of terms) the corresponding matched filter is, as we have already seen, easily realised by a normal type of non-recursive (transversal) digital filtering operation. To summarise, the performance of a matched filter represents an upper bound to the improvement in signal-to-noise ratio which may be achieved by linear filtering, prior to detection of a known signal waveform in the presence of additive noise. When dealing with continuous signals and analogue filters, we are forced to settle for a filter performance rather less than the optimum.

10.3.3 Pulse compression techniques

Having dealt with the theoretical performance of the matched filter, we now consider a further important practical aspect of this signal processing technique. This is generally referred to as pulse compression. Suppose, for example, it is decided to design a pulse radar system in which each transmitted pulse is spread over a considerably longer time-interval than the normal 1 microsecond or so. This has the advantage of reducing the *peak* transmitter power required, while maintaining the total energy contained in each pulse (and, as we have seen, the detectability of a known waveform in noise depends on its total energy rather than its detailed waveshape). The main disadvantage is that a stretched radio frequency pulse gives poor range resolution (since radio waves travel a go-and-return path of 1 km in about 6 μs, an echo of this duration represents a 'blip' about 1 km across). An effective solution is to code the stretched transmitter pulse with a wideband modulating signal, and use a matched filter to sharpen up, or compress, the echo again in the receiver.

The general principle is summarised by figure 10.12, using a PRBS in place of the conventional rectangular pulse. As we saw in section 5.5.4, a PRBS is a two-level

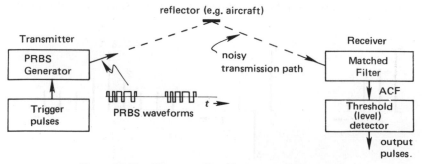

Figure 10.12   *Elements of a pulse compression system*

pseudo-random signal which autocorrelates to a narrow spike at $\tau = 0$. If such a waveform is transmitted, and passed through its matched filter in the receiver, it will therefore compress into a narrow pulse of width equal to one PRBS clock period. Note that the PRBS can be of any length, without affecting timing resolution in the receiver. Thus signal duration can, in principle, be traded against the peak power level required of the transmitter. In the radar example outlined above, the PRBS would, of course, be used to modulate the radio frequency wave put out by the transmitter; but in other cases a high-frequency carrier may not be involved, and the PRBS could be transmitted directly. There are other potential advantages in using such a technique in a variety of applications. Very narrow pulses are sometimes lost during transmission — for example, if the signal channel is subject to short noise bursts — and it may be helpful to spread them over a longer time-interval. Secondly, the transmission of a low-level noise-like signal may offer security advantages. And finally, the use of an extended low-level pulse in place of a short, high-level one may avoid overloading the transmission system, or driving it into non-linearity.

There is, of course, no need to use a PRBS; a Huffman sequence would also serve. Nor does the code have to be pseudo-random: quite often a swept-frequency pulse, a so-called 'chirp' waveform, is used. The main requirement is that the transmitted code is sufficiently wideband to give a sharply-defined autocorrelation peak in the receiver.

Figure 10.13 illustrates some typical waveforms. Part (a) shows a 31-character PRBS, represented as a sampled-date signal with values ±1. Part (b) shows its ACF, to a different vertical scale. (Note that there are small, variable residues to either side of the peak at $\tau = 0$, whereas the ACF previously shown in figure 5.19 indicates a constant negative residue. This is because we earlier assumed a continuously repeating PRBS, whereas we are now dealing with the transmission and reception of a *single* waveform. We might also note, in passing, that a single so-called Barker code offers the advantage of small, *constant* amplitude, residues, and is therefore preferable in this respect to a PRBS. However, Barker codes have a maximum length of only 13 binary characters.[21] The triangular form of the ACF peak in figure 5.19 is due of course, to the assumption of a square-pulse format for the PRBS, rather than the present sampled-data format.) Figure 10.13(c) shows sampled white gaussian noise of unit variance, into which are inserted three repetitions of the PRBS at positions indicated by the horizontal bars. This represents three signal waveforms so severely corrupted by noise that it is impossible to tell, by inspection, where they occur. Part (d) of the figure shows the signal-plus-noise waveform after matched filtering. The signal autocorrelation peaks, indicated by arrows, stand out very clearly — even though two of the PRBS waveforms overlapped on the input side. Using the theoretical result already derived, the improvement in S:N ratio in this case is readily calculated as $\sqrt{31} = 5.6$. Finally, it is worth noting that a PRBS offers the best possible improvement in S:N ratio for a given duration of signal, because all its samples are equal in absolute value. This gives the largest possible ratio between total energy, and peak value prior to filtering.

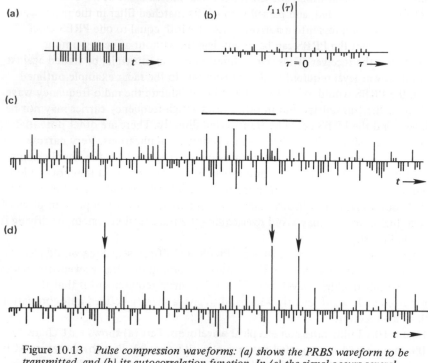

Figure 10.13   *Pulse compression waveforms: (a) shows the PRBS waveform to be transmitted, and (b) its autocorrelation function. In (c) the signal occurs several times, severely contaminated by wideband noise; (d) shows the matched filter output*

It is sometimes asked why a waveform such as that of figure 10.13(d) cannot be further improved in S : N ratio by another matched filtering operation. After all, we know the new form of the 'signal', and it is still corrupted by noise. The answer is that no further improvement is possible, because the signal components have been optimally arranged in phase by the first matched filter to yield the largest possible peak, and the signal and noise at the output of such a filter always have the same spectral distribution. Thus another linear filter, which could only adjust spectral amplitudes and/or phases, could offer no advantages.

10.3.4 Detection errors

When a signal-plus-noise waveform has been filtered as effectively as possible — quite often by using a matched filter — there is still a final decision to make about where the signal waveform occurs. This is a statistical problem which cannot be answered with certainty: even if a particular feature in a waveform looks like a signal peak, it is at least possible that it is only noise. In other words, we must be prepared to accept some probability of errors of judgement — even when that probability has been substantially reduced by initial filtering.

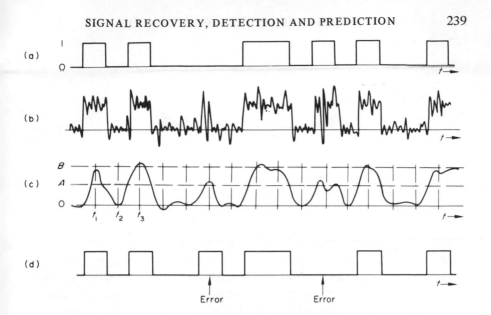

Figure 10.14 *Typical waveforms in a pulse-code communication system:*
(a) transmitted pulse-code, (b) received signal, (c) filtered signal, and (d) final
detected pulse-code

Figure 10.14 illustrates an important example. Part (a) represents a transmitted binary signal in a pulse-code communication system: upon reception, the signal (b) has been substantially degraded by noise. In part (c) suitable filtering has enhanced the ratio between peak signal and standard deviation of the noise. Suppose we now make a decision about the presence or absence of a transmitted pulse by using a threshold level A: if the signal exceeds A at instants $t_1, t_2, t_3, \ldots$, we will assume a pulse to have been transmitted; but if A is not reached, we will assume no pulse to have been transmitted. In this situation, two types of error are possible:

(a) no pulse is sent, but the noise is itself sufficient to exceed the threshold, and
(b) a pulse is sent, but the noise is so negative that the threshold is not reached.

One error of each type is shown in figure 10.14(d). The probability of error is, of course, crucially dependent upon the level of the noise. If the noise amplitude distribution is symmetrical in form and has zero mean value, it is intuitively clear that the two above types of error will be equiprobable when the decision threshold A is equal to half the peak value B reached by the pulse signal alone.

The error probability may be assessed if the noise amplitude distribution is known. In figure 10.15, the noise is assumed gaussian with zero mean level: part (a) shows its amplitude probability density distribution, which is, of course, the same as that of the signal-plus-noise waveform when no pulse is present; part (b) of the figure shows the density distribution relevant to detection instants when the pulse is present at level B. The shaded areas give the probabilities of occurrence of the two types of error mentioned above: in the case of gaussian noise, these areas are

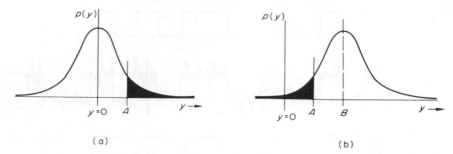

Figure 10.15 *Amplitude probability density functions for the waveform of figure 10.14(c): (a) noise alone and (b) noise plus signal at level B*

tabulated as the so-called error (erf) function in many texts on statistics. It is interesting to note that when the noise is gaussian, the error probabilities decrease quite dramatically with noise level because of the very rapid decay of the 'tails' of the gaussian distribution (see section 5.5.2). For example, if the two received signal levels 0 and B are equally likely to occur and the detection threshold A is set to $B/2$, the error probability is related to the standard deviation of the noise as follows[5]

B/σ	error probability
4.6	10^{-2}
7.4	10^{-4}
9.4	10^{-6}
11.2	10^{-8}

Suppose we have a pulse-code modulation system (see section 8.3.3) for speech transmission, in which a speech waveform is sampled 8000 times per second and 7 binary characters are allocated to the coding of each sample value: 56 000 characters must therefore be transmitted per second. If the noise level at the receiver, after suitable filtering, is such as to give $B/\sigma = 9.4$, there will be on average $56\,000 \times 10^{-6} = 0.056$ errors/second — or one error every 18 seconds — which is almost certainly quite acceptable. However, if the noise level at the receiver rises by as little as 27%, giving $B/\sigma = 7.4$, we get an error, on average, every 0.18 seconds — which would be quite unacceptable. The relatively narrow range of signal-to-noise ratios over which the error rate changes from negligible to serious levels gives rise to the term 'threshold effect' (which must not be confused with the use of a detection threshold such as level A in figures 10.14 and 10.15).

We have chosen a very simple case where the signal to be detected takes on only two possible values. However, the same approach is relevant to a multi-valued signal waveform, and allows the probabilities associated with the various types of error to be estimated. There are, once again, no differences in principle involved whether we are dealing with sampled-data or continuous signals.

10.4 Signal prediction

10.4.1 Introduction

In previous sections we have dealt with the recovery and detection of signals contaminated by random noise. We now turn to the rather different problem of trying to predict the future of a random signal on the basis of its present value and past history. As we shall see, this problem is naturally tackled in the time-domain, and has close links with the techniques of optimum signal estimation discussed in section 10.2.4.

Let us consider more carefully the example of trying to predict the future position of an aircraft, already mentioned in the introduction to this chapter. Suppose the pilot is trying to take evasive action, which means that he is trying to be as *unpredictable* as possible. However, he is subject to various constraints: the dynamics of the aircraft (how quickly it responds to sudden changes of the flight controls, and so on); the maximum stresses he can safely impose on the aircraft structure; and the maximum accelerations the human body can withstand. All these factors make the future position of the aircraft, in fact, partly predictable. In signal processing terminology, we might say that the aircraft's position, considered as a random signal, is not like white noise; but successive values of that signal are to some extent correlated. It is this correlation, or statistical *structure*, which must be used in any attempt to predict its future. Such prediction would be needed by an anti-aircraft gunner, who must not aim where the aircraft is *now*. Rather, he is concerned to bring his shell and the aircraft together at some point in space in the *future*. It was indeed this problem, made more difficult for the unaided human operator by the ever-increasing speed of aircraft, which stimulated some of the earliest theoretical work into automatic prediction systems during the Second World War.

If a linear time-invariant filter is to be designed for automatic prediction, the statistical properties of its input signal must of course be known, or assumed, in advance. Furthermore, those statistics must be stationary. Both conditions make the practical design of a predictor for aircraft tracking much more difficult than the above discussion might suggest. Nevertheless it serves as a good example of the type of problem covered in this final section.

It is perhaps natural to suppose that a random signal will be fairly predictable only if it is varying rather slowly — as in our aircraft example, where the constraints mentioned will certainly tend to produce a signal of predominantly low-frequency content. At this point it is however helpful to return to the sampled-data signals shown in figure 10.9. Part (a) shows a portion of white noise, truly unpredictable because knowledge of its present and previous values is no help whatsoever in deciding what the next value(s) will be. Parts (b) and (c) show signals increasingly strong in low frequencies, and it is clear that (c) would be fairly predictable because it tends to change rather little from one value to the next. Note, however, that the waveform of figure 10.9(d) would be just as predictable as that in (c), even though it is varying very rapidly. We can be almost certain that the next value will be opposite in sign to the present one, and so on. What matters is not which frequencies are dominant, but how tightly the spectrum is constrained. Bandwidth is therefore the important factor. A narrow frequency range implies an autocorrelation function which spreads considerably around $\tau = 0$. This is the time-domain effect which we need for successful prediction.

10.4.2 The Wiener predictor

The design of automatic predictors follows closely the arguments developed in section 10.2.4 for optimum signal estimation in the presence of noise. Given the present value, past history, and spectral or autocorrelation properties of a random signal, we may define either a Wiener or a Kalman filter which gives, on average, the best possible prediction of the next value of the signal in a least-squares sense. With *estimation*, the 'true' signal is always corrupted by noise (if it were not, an estimating filter would not be needed). But in the case of *prediction*, two situations may arise: (a) the present and previous values of the signal are assumed to have been measured accurately, or (b) they are corrupted by measurement noise. Case (b) is more general, but for the purposes of this introduction to automatic prediction we will consider only the simpler case (a). Furthermore, we will illustrate some of the main aspects of prediction by reference to just the Wiener technique. Interested readers will find both Wiener and Kalman predictors covered more fully in several recent texts (for example, Bozic,[41] Gelb[48]).

The Wiener–Hopf equation, derived and discussed in section 10.2.4.2, takes the form

$$\sum_{i=0}^{k} a_i r_{11}(i, j) = r_{12}(j) \quad \text{for} \quad j = 0, 1, 2, \ldots, k$$

where r_{11} is the autocorrelation function (ACF) of the input to an optimum filter, a_i are the non-recursive filter coefficients, and r_{12} is the desired crosscorrelation function (CCF) between the filter input and output. The same equation may readily be shown to apply to optimum prediction: r_{11} is now the ACF of the random signal to be predicted; and since the *desired* output signal from a predictor is a time-advanced version of its input signal, r_{12} must be identical to r_{11} apart from a time-shift.

Let us illustrate with an example. Figure 10.16 shows a random signal generator, followed by a predictor whose output y is required to be a time-advanced version of its input x. The prediction will not be perfect, and the instantaneous error e is found by comparing x with a delayed version of y. We will work with sampled-data signals, and assume that prediction by one sampling interval T is required. (It would, of course, be possible to predict further forward than this, at the expense of a greater mean square error.)

The random signal generator consists of a zero-mean white noise source of unit variance, followed by a filter to shape its spectral characteristics and limit its bandwidth. In general, a filter will also modify the level of the signal, and the variance of its output will be some other value, say c. For convenience, the signal delivered to the predictor will be converted back to unit variance by including a

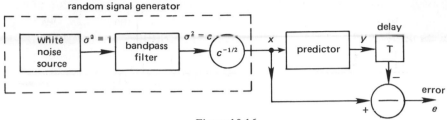

Figure 10.16

multiplier coefficient $c^{-1/2}$. For illustration purposes, a wide variety of signal spectral characteristics could be used. Here we choose a bandpass filter with an impulse response $I_1(t)$ having the following terms: 4, 8, 4, −3, −6, −3, 2, 4, 2, −1, −2, −1. This is shown in figure 10.17(a). Note that since $I_1(t)$ is oscillatory at about 6 samples per cycle, we must expect the resulting random signal to be strong in frequency components close to $1/6T$ Hz, or $\omega = \pi/3T$ radians per second.

The ACF of the filter output is important for two reasons: firstly, because we need it to design the predictor; and secondly, because by taking its Fourier transform we can find the power spectrum of our random signal. As already noted in the discussion of random signal processing in section 7.4, and summarised by figure 7.20, the ACF of any linear processor's output signal equals its input ACF convoluted with the ACF of its impulse response. In this case we have a unit-variance white noise input, which autocorrelates to a unit-valued 'spike' at $\tau = 0$; so the ACF of the output signal will be just that of the impulse response $I_1(t)$. This is shown in figure 10.17(b). Its peak value (equal to the sum of the squares of the terms in $I_1(t)$) is 180 and it is, of course, symmetrical about $\tau = 0$. Its values, for positive τ, are: (180), 100, −50, −120, −69, 24, 66, 40, −5, −24, −16, −4. The Fourier transform of this ACF, which equals the power spectrum $P_{11}(\omega)$ of the random signal, has the form shown in part (c) of the figure. It clearly demonstrates the bandpass action of the filter and, as expected, peaks at about $\omega = \pi/3T$. Note that the signal's variance is 180, so that the multiplier shown in figure 10.16 is set to $180^{-1/2} = 0.0745$. The random signal delivered to the predictor will now have unit variance, and an ACF the same as that of figure 10.17(b) but normalised to a peak value of unity.

Having established the statistical properties of our random signal, we now turn to the design of the predictor. As already noted, we will use the non-recursive Wiener approach, and must first decide how many impulse response coefficients to specify. Since the ACF of the signal to be predicted spreads to $\tau = \pm 11T$, we may expect that a predictor with about 11 coefficients will give a good performance. Let us choose 10. The Wiener−Hopf equation then gives us 10 simultaneous algebraic equations to solve, of which the first and the last are

$$r_{11}(0) \cdot a_0 + r_{11}(1) \cdot a_1 + \ldots + r_{11}(8) \cdot a_8 + r_{11}(9) \cdot a_9 = r_{12}(0) = r_{11}(1)$$

$$\vdots$$

$$r_{11}(9) \cdot a_0 + r_{11}(8) \cdot a_1 + \ldots + r_{11}(1) \cdot a_8 + r_{11}(0) \cdot a_9 = r_{12}(9) = r_{11}(10)$$

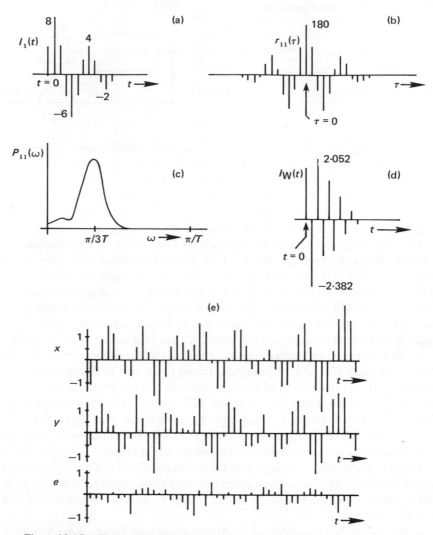

Figure 10.17   *Illustration of a Wiener predictor: (a) impulse response of the filter used to produce the random signal, (b) its ACF, (c) the power spectrum of the signal, and (d) impulse response of the 10-term Wiener predictor. In (e) are shown typical portions of the signal, predictor output, and error*

Note that $r_{12}(0) = r_{11}(1)$ and $r_{12}(9) = r_{11}(10)$ because the desired input–output CCF is identical to the input ACF but time-shifted by one sampling period. We know all the required ACF values, which are as quoted above — but divided by 180. The normal method of solving such a set of equations is by matrix inversion on a digital computer. This yields the following values for the predictor coefficients

$$a_0 = \ \ 1{\cdot}7885 \qquad a_4 = \ \ 1{\cdot}3587 \qquad a_7 = -0{\cdot}4870$$

$$a_1 = -2{\cdot}3820 \qquad a_5 = -1{\cdot}0774 \qquad a_8 = \ \ 0{\cdot}2683$$

$$a_2 = \ \ 2{\cdot}0521 \qquad a_6 = \ \ 0{\cdot}7846 \qquad a_9 = -0{\cdot}1013$$

$$a_3 = -1{\cdot}7092$$

These coefficients define the impulse response $I_W(t)$ of the Wiener predictor, which is shown in figure 10.17(d).

The choice of ten Wiener coefficients seems to have been sensible because the last few of them are quite small. Since successive values alternate in sign, the predictor must have high-pass frequency response characteristics. Why this should be so, when it is required to predict a bandpass signal, is not easy to explain: indeed it seems best not to concern ourselves unduly with frequency-domain properties, when the predictor has been designed using time-domain criteria. One interesting general point is that an ideal predictor, valid for *any* signal, would have an impulse response consisting of a single, unit-height sample value occurring before $t = 0$ (corresponding to the required prediction time – in our example, at $t = -T$). But such an impulse response is not physically realisable. What the Wiener design technique yields is the optimum realisable predictor for a *particular* signal, for a stated number of impulse response coefficients. It nevertheless remains hard to see how the $I_W(t)$ shown in figure 10.17(d) can have a similar effect to the ideal predictor, for the random signal we have chosen. About all we can say is that our ten-coefficient design produces correct values of the desired input–output CCF at ten positive values of τ, but errors elsewhere. The ideal predictor would, of course, achieve the desired CCF exactly.

A typical portion of the random signal (x) is shown together with the predictor's output (y) in figure 10.17(e). Careful scrutiny reveals that each output value is indeed a reasonably successful prediction of the *next* input. If the average size of the prediction error, plotted below, seems disappointing, it is important to remember that the power spectrum $P_{11}(\omega)$ of our random signal is not very tightly constrained, so that the signal is not, in fact, very predictable. Even the best linear predictor cannot achieve the impossible.

Although we will omit a detailed analysis of prediction error, for reasons of space, the nature of this error is worth further consideration. We have implied above that since the last few of our Wiener coefficients are quite small, there would have been little benefit in specifying a larger number of them. In other words, the error shown at the bottom of figure 10.17 is probably about as small, on average, as could be achieved with any realisable linear predictor. One way of converting this intuitive argument into a quantitative one is to examine the spectral (or auto-correlation) characteristics of the error. If the error is white (successive values totally uncorrelated), it cannot contain any further information about the signal we are trying to predict. Or, to put the argument the other way round, if the error is not white, knowledge of its present and previous values must allow us to predict, to some extent, its next value – and we could use this information to reduce the error and thereby improve our prediction of the signal. Hence a test for the whiteness of the error is often used to indicate whether or not the prediction could be improved.

Closely related to the above argument is the idea of using linear predictors for modelling random signals and processes. This technique has found many recent applications, from the synthesis of speech[49] to the modelling of electrical activity in human muscles.[50] Although the topic is too complex to be described fully here, its basis may quite easily be summarised by referring back to figure 10.16. In that figure, the system to the right-hand side of the random signal generator (comprising predictor, delay, and subtractor) may be viewed as a linear filter which converts a random signal x into an error 'signal' e. As we have seen, e is white if the prediction is the best that can be achieved. If this linear system for converting x into e is described by a transfer function $G(z)$, then it is clear that the inverse function $H(z) = G(z)^{-1}$ would transform white noise into a signal with the statistical properties of x. But this is exactly what our random signal generator on the left-hand side of figure 10.16 does. Therefore if we can devise (by whatever method) an optimum predictor for a random signal, we may derive from it a transfer function $H(z)$ which can serve as a model of the process generating that signal. This possibility greatly increases the practical importance of the techniques of linear prediction.

Problems

1. A signal occupies the frequency range $-\omega_0 < \omega < \omega_0$ and is contaminated by random noise having energy in the range $-3\omega_0 < \omega < 3\omega_0$. Estimate the improvement in the ratio of peak signal to r.m.s. noise attainable with an ideal recovery filter, when the noise power spectral density characteristic is

   (i) constant over the above range;
   (ii) triangular in form with its peak value at $\omega = 0$.

2. A signal has significant frequency components in the range $-400\pi < \omega < 400\pi$ radians/second, and is degraded by a narrowband interference at mains supply frequency (50 Hz, or 100π radians/second). It is decided to filter out the interference using a digital filter, after sampling at a rate of 500 samples/second. The filter's transfer function is

$$H(z) = \frac{\{z - \exp(j\omega_0 T)\}\{z - \exp(-j\omega_0 T)\}}{\{z - r.\exp(j\omega_0 T)\}\{z - r.\exp(-j\omega_0 T)\}},$$

   where

$$r = 0.990,$$

$$\omega_0 = 100\pi$$

   and T is the sampling interval.
   Draw the pole-zero configuration of the filter, and sketch its frequency

response characteristic, in both magnitude and phase, in the range $0 < \omega < \pi/T$. What is the advantage of including the complex conjugate pole pair? Finally, evaluate the recurrence formula of the filter.

3. The power spectrum of a stationary random sampled-data signal is given by

$$P(\omega) = 4 + 6 \cos \omega T + 4 \cos 2\omega T + 2 \cos 3\omega T$$

It is contaminated by sampled white noise having zero mean and unit variance. Find the coefficients of the optimum 3-term Wiener filter for estimating the true value of the signal.

4. A signal with successive sample values 5, −4, 3, −2, 1 is to be detected in the presence of wideband noise by use of a digital matched filter.

   (i) Define the filter's impulse response;
   (ii) What is the filter's output due to the signal alone?
   (iii) What improvement will the filter give in the ratio of peak signal to standard deviation of the noise?
   (iv) Compare the matched filter's performance with that of a filter whose impulse response is identical to (and not a time-reversed version of) the signal itself. What is the relationship between the frequency responses of the two filters, and why is the matched filter more effective?

5. Figure 10.14 in the main text illustrates the detection of a binary pulse signal in the presence of noise. Assume the noise to be gaussian with zero mean and a standard deviation σ, after initial filtering. By consulting a table of the error function (erf), find the probabilities of the two types of detection error illustrated, when the threshold level A is set to half the peak signal level B, and

   (i) $\sigma = B/6$,
   (ii) $\sigma = B/8$.

What is the effect on these results of making $A = B/3$?

6. Sampled white noise is processed by a high-pass digital filter whose impulse response has the following values: 1, −1, 1, −1, 1, −1. At the filter output, the noise (now considered as a random signal) has unit variance. Assuming you have access to a digital computer with a matrix inversion routine in its standard software, find the impulse response of the eight-term Wiener predictor for predicting the value of the signal

   (a) one sample ahead, and
   (b) two samples ahead.

You may also wish to check its performance by simulation on the computer, using a random number generator.

Postscript

In this book, we have been concerned above all with the constant interplay
between time and frequency domain descriptions of signals and systems, which is
perhaps the main hallmark of signal analysis and linear signal processing. An
attempt has been made to fit random waveforms into the same general framework as deterministic ones, and to show how the effects of linear processing on
both types of waveform may be assessed. We have only rarely encountered
functions—such as the amplitude distribution of a signal—which fail to fit
naturally into this general theme. The reasons for the emphasis on time and
frequency descriptions are really twofold: on the one hand, linear processors
form a class of device which has found very widespread practical application;
and the design and development of linear systems has been greatly stimulated
by the relative simplicity of their mathematical description. But in spite of these
advantages of linear processing, the reader is asked to bear in mind that a variety
of useful signal techniques may be achieved by using nonlinear systems and
devices, and that the analysis and synthesis of signals and systems by reference
to Fourier and Laplace transform techniques is only part of the story. This
general point has been made several times in the text, for example by discussing
classes of nonsinusoidal orthogonal functions in chapter 2, and by making a
brief reference to nonlinear processing at the end of chapter 7.

Considerable attention has been paid to sampled-data signals. Once again,
there are two main reasons for this. Firstly, a number of important concepts—
such as convolution and matched filtering—are really quite straightforward
when discussed by reference to sampled-data signals, whereas the conventional
treatment in terms of continuous signals and analogue systems tends to involve
much more demanding mathematics. Secondly, there is little doubt that sampled
signals have arrived and that they will continue to grow in practical importance.
The more signals and data are stored in digital computers, and transmitted from
place to place in the form of discrete electrical pulses, the greater will become the
incentive to process them in sampled-data form. The trend towards digital
signal processing must be expected to continue.

APPENDIX: Some Useful Laplace Transforms

$f(t)$, for $t > 0$	Waveform	$G(s)$	S-plane poles and zeros
$\delta(t)$		1	
1		$\dfrac{1}{S}$	
t		$\dfrac{1}{S^2}$	
$e^{-\alpha t}$		$\dfrac{1}{(S+\alpha)}$	$S = -\alpha$
$(1-e^{-\alpha t})$		$\dfrac{\alpha}{S(S+\alpha)}$	$S = -\alpha$
$(e^{s_1 t} - e^{s_2 t})$		$\dfrac{S_1 - S_2}{(S-S_1)(S-S_2)}$	
$\sin \omega_0 t$		$\dfrac{\omega_0}{S^2 + \omega_0{}^2}$	
$\cos \omega_0 t$		$\dfrac{S}{S^2 + \omega_0{}^2}$	
$e^{-\alpha t}\sin \omega_0 t$		$\dfrac{\omega_0}{(S+\alpha+j\omega_0)(S+\alpha-j\omega_0)}$	
$\dfrac{\sin(\omega_0 t-\theta)+e^{-\alpha t}\sin \theta}{(\alpha^2 + \omega_0{}^2)^{1/2}}$ where $\theta = \tan^{-1}\dfrac{\omega_0}{\alpha}$		$\dfrac{\omega_0}{(S+\alpha)(S^2 + \omega_0{}^2)}$	

Bibliography

(A) General books which deal largely, or exclusively, with continuous (analogue) signals and systems

(1) B. P. Lathi. *An introduction to random signals and communication theory.* Intertext, London (1970)
(2) B. P. Lathi. *Signals, systems and communication.* Wiley, New York (1967)
(3) F. G. Stremler. *Introduction to communication systems.* Addison Wesley, Reading, Mass. (1977)
(4) Y. W. Lee. *Statistical theory of communication.* Wiley, New York (1960)
(5) M. Schwartz. *Information transmission, modulation and noise.* McGraw-Hill, New York (1970)
(6) H. Taub and D. L. Schilling. *Principles of communication systems.* McGraw-Hill, New York (1971)
(7) R. W. Coates. *Modern communication systems.* Macmillan, London (1975).
(8) H. H. Skilling. *Electrical engineering circuits.* Wiley, New York (1965)

(B) Books which cover the z-transform and sampled-data signals and/or systems

(9) S. G. Gupta. *Transform and state variable methods in linear systems.* Wiley, New York (1966)
(10) A. P. Oppenheim and R. W. Schafer. *Digital signal processing.* Prentice-Hall, Englewood Cliffs, N.J. (1975)
(11) L. R. Rabiner and B. Gold. *Theory and application of digital signal processing.* Prentice-Hall, Englewood Cliffs, N.J. (1975)
(12) B. Gold and C. M. Rader. *Digital processing of signals.* McGraw-Hill, New York (1969)

(C) Introductory books in probability and statistics

(13) C. Mack. *Essentials of statistics for scientists and technologists.* Heinemann, London (1966)
(14) M. Goldman. *Introduction to probability and statistics.* Harcourt, Brace and World, New York (1970)
(15) W. B. Davenport. *Probability and random processes.* McGraw-Hill, New York (1970)

(D) More specific references

(16) D. C. Champeney. *Fourier transforms and their physical applications.* Academic Press, London (1973)
(17) J. L. Stewart. *Fundamentals of signal theory.* McGraw-Hill, New York (1960)
(18) H. K. Crowder and S. W. McCuskey. *Topics in higher analysis.* Macmillan, New York (1964)

(19) J. L. Walsh. A closed set of normal orthogonal functions. *Ann. Math.*, **45**, (1923) 5

(20) W. T. Cochran et. al. What is the Fast Fourier Transform? *Proc. I.E.E.E.*, **55** (1967) 1664

(21) G. Hoffmann de Visme. *Binary sequences.* English Universities Press, London (1971)

(22) D. A. Huffman. The generation of impulse-equivalent pulse trains. *I.R.E. Trans. Inf. Theory.*, **8** (1962), p.S. 10

(23) D. R. Cox and V. Isham. *Point processes.* Chapman & Hall, London (1980)

(24) D. H. Perkel et. al. Neuronal spike trains and stochastic point processes. *Biophys. J.*, **7** (1967), 319

(25) D. Graupe. *Identification of systems.* (2nd edition). Krieger, New York (1976)

(26) M. H. Ackroyd. *Digital filters.* Butterworth, London (1973)

(27) C. M. Rader and B. Gold. Digital filter design techniques in the frequency domain. *Proc. I.E.E.E.*, **55** (1967), 149

(28) P. A. Lynn. Recursive digital filters with linear phase characteristics. *Comput. J.*, **15** (1973), 337

(29) M. G. Kendall and A. Stuart. *The advanced theory of statistics*, Vol. I. Griffin, London (1963)

(30) E. Kreyszig. *Introductory mathematical statistics.* Wiley, New York (1970)

(31) J. L. Marshall. *Introduction to signal theory.* Intertext, New York (1965)

(32) R. B. Blackman and J. W. Tukey. *The measurement of power spectra.* Dover, New York (1959)

(33) M. S. Roden. *Analog and digital communication systems.* Prentice-Hall, Englewood Cliffs (1979)

(34) J. D. Rhodes. *Theory of electric filters.* Wiley, London (1976)

(35) P. Bowron and F. W. Stephenson. *Active filters for communication and instrumentation.* McGraw-Hill, London (1979)

(36) E. Williams. *Electric filter circuits.* Pitman, London (1963)

(37) G. B. Clayton. *Operational amplifiers.* Butterworth, London (1971)

(38) F. R. Connor. *Networks.* Arnold, London (1972)

(39) W. E. Thomson, Networks with maximally flat delay. *Wireless Engr*, **29** (1952), 256

(40) P. A. Lynn, Recursive digital filters for biological signals. *Med. biol. eng.*, **9** (1971), 37

(41) S. M. Bozic. *Digital and Kalman filtering.* Arnold, London (1979)

(42) P. A. Lynn. 'Economic, linear phase, recursive digital filters'. *Electron. Lett.*, **6** (1970) 143

(43) P. A. Lynn. 'On-line digital filters for biological signals: some fast designs for a small computer'. *Med. Biol. Engng Comput.*, **15** (1977) 534

(44) P. J. van Gerwen, W. F. G. Mecklenbraucker, N. A. M. Verhoeckx, F. A. M. Snijders and H. A. van Essen. 'A new type of digital filter for data transmission'. *I.E.E.E. Trans.*, **COM–23** (1975) 222

(45) P. A. Lynn. 'FIR digital filters based on difference coefficients: design

improvements and software implementation'. *I.E.E. Proc.*, **127**, Part E (1980) 253

(46) T. J. Terrell. *Introduction to digital filters*. Macmillan, London (1980)

(47) B. S. Tan and G. J. Hawkins. 'Speed-optimised microprocessor implementation of a digital filter'. *I.E.E. Proc.*, **128**, Part E (1981) 85

(48) A. Gelb. *Applied optimal estimation*. MIT Press, Cambridge, Mass. (1974)

(49) J. D. Markel and A. H. Gray. *Linear prediction of speech*. Springer-Verlag, New York (1976)

(50) D. Graupe, J. Magnusson and A. A. Beex. 'A microprocessor system for multifunctional control of upper limb prosthesis via myoelectric signal identification'. *I.E.E.E. Trans.*, **AC–23** (1978) 538

Index